Wheelwright
Protein Purification

Scott M. Wheelwright

Protein Purification: Design and Scale up of Downstream Processing

with 53 Figures and 39 Tables

Hanser Publishers, Munich Vienna New York Barcelona

Distributed in the United States of America and in Canada by
Oxford University Press, New York

Author: *Dr. Scott M. Wheelwright* CHIRON Corporation Emeryville CA 94608 USA

Distributed in USA and Canada by
Oxford University Press
200 Madison Avenue, New York, N.Y. 10016

Distributed in all other countries by
Carl Hanser Verlag
Kolbergerstraße 22
D-8000 München 80

Library of Congress Cataloging-in-Publication Data

Wheelwright, Scott M.
Protein purification : design and scale up of downstream processing / Scott M. Wheelwright.
p. cm.
Includes index.
ISBN 0-19-520913-3
1. Proteins – Biotechnology. 2. Proteins – Purification.
I. Title.
TP248.65.P76W44 1991 91-71911
660'.63–dc20 CIP

Die Deutsche Bibliothek – CIP-Einheitsaufnahme

Wheelwright, Scott M.:
Protein purification : design and scale up of downstream processing / Scott M. Wheelwright. – Munich ; Vienna ; New York ; Barcelona : Hanser, 1991
ISBN 3-446-15703-4

ISBN 3-446-15703-4 Carl Hanser Verlag Munich Vienna New York Barcelona
ISBN 0-19-520913-3 Oxford University Press

Library of Congress Catalog Card Number: 91-071911

cover photo credit

Cover photo courtesy of Jeff Blaney, Protos Corporation. Insulin structure 4INS from Brookhaven Protein Data Bank with A *chain in red and* B *chain in yellow displayed by MIDAS' cartoon display program.*

Brookhaven Protein Data Bank
Bernstein, F. C., T. F. Koetzle, G. T. B. Williams, E. F. Meyer, M. D. Brice, J. R. Rodgers, O. Kennard, T. Shimanouchi, and M. Tasumi, *Journal of Molecular Biology* 112:535 (1977).

MIDAS
Ferrin, T. E., C. C. Huang, L. E. Jarvis, and R. Langridge, *Journal of Molecular Graphics* 6:13 (1988).

Cover design: Kaselow Design, Munich
Printing and binding: Passavia Druckerei GmbH Passau
Printed in Germany

PREFACE

One of the dominant characteristics of high–technology industries is the overlap of skills needed between disciplines. The boundaries between fields of study are no longer sharp and distinct but rather are blurred to a considerable extent. Biotechnology is not an exception. For products to move from the laboratory to commercial success requires more than just an understanding of the new biology; it also requires an understanding of the fundamentals of engineering and their practice as applied to large–scale protein purification.

The roles of the biochemist and the chemical engineer have been traditionally distinct, but an overlap between these two disciplines has been created by biotechnology. The number of proteins that have been identified as commercial products has generated a need for scientists and technicians with experience in both areas. The purpose of this book is to help bridge the historical gap, to provide the biochemist with an understanding of engineering requirements and the engineer with an understanding of biochemistry requirements for successful scale up of processes for the purification of proteins on a large scale and the commercial success of the new biology.

The hybridization of the biochemist or protein chemist and the chemical engineer or process engineer creates a new discipline, that of protein process designer. This new discipline incorporates both the protein chemist's understanding of protein chemistry—structure, function, characterization, and reaction—and process engineering: mass transfer, hydrodynamics, and thermodynamics. And although the fundamentals underlying this discipline are held in common with other fields, the applications are unique. It is the application of the fundamentals that necessitates this welding of disciplines into a new field.

This book concentrates on these unique applications; for it is the applications that truly set this area apart from other fields. And it is the sheer volume of possible applications or unit operations for protein purification that requires a specialist in this area. Gone are the days when a generalist with an understanding of the fundamentals could pick up and make use of each of these applications without further training. And while it is still possible to intuit each of these applications, starting with first principles, to do so would require a lifetime. Thus this book has been written to provide an introduction to the art and science utilized by the biochemist and process engineer in the design and optimization of large–scale protein–purification processes.

The author wishes to acknowledge the help of many who have brought this work to fruition: my family for their love and support, my colleagues, especially Dr. Ron Johnson, for their encouragement, ideas, suggestions, and helpful pointers on large–scale protein purification, and Scott Bevan for his fine drawings. Parts of this book were previously published: Chapter 2 in the *Journal of Biotechnology* 11 (1989), and Chapter 5 in Biotechnology Research and Applications, J. Gavora et al. (eds.), Elsevier Applied Science, London (1988).

CONTENTS

NOMENCLATURE

A	=	cross–sectional area
$b_p(G_{sol})$	=	free energy resulting from interaction of salts and bound protein
c	=	solute concentration in fluid phase
c_b	=	concentration in bottom phase
c_f	=	final concentration
c_i	=	initial concentration
c_r	=	reference concentration in fluid phase
c_t	=	concentration in top phase
C_d	=	drag coefficient
C_0	=	total ionic strength of solution
d_p	=	diameter of particle
D	=	diffusivity
$f_p(G_{sol})$	=	free energy resulting from interaction of salts and unbound protein
F_b	=	buoyancy force
F_c	=	centrifugal force
F_d	=	drag force
g	=	gravitational acceleration
h	=	reduced plate height
$HETP$	=	height equivalent to a theoretical plate
J	=	flux
k	=	valence ratio of ions
K	=	equilibrium constant
K_s	=	salting–out constant
L	=	length of column
m	=	mass
M_0	=	zero–order moment
M_1	=	first–order moment
M_2	=	second–order moment
M_i	=	ith–order moment
n	=	exponent in Stokes–Einstein equation
N	=	speed of rotation (number of revolutions per unit time)
NTP	=	number of theoretical plates
P	=	transmembrane pressure
P_f	=	filtrate pressure
P_i	=	inlet pressure
P_o	=	outlet pressure
q	=	solute concentration in solid phase
Q	=	volumetric flow rate
Q_0	=	total number of charge sites in solid phase
r	=	radius
r_1	=	outer radius of centrifuge rotor

r_2 = inner radius of centrifuge rotor
R = constant separation factor
R_g = resistance of gel layer
R_m = resistance of membrane
RCF = relative centrifugal force
S = sedimentation coefficient
t = time
v = volume of sorbent from inlet to point of interest
V = volume of fluid fed to column
V_c = cumulative volume
V_e = elution volume
V_f = final volume
V_i = volume injected
V_s = volume of stationary phase
V_t = total volume of column
V_v = void volume of column
V_0 = average sample volume
V' = holdup volume
V_0' = volume after concentration and dilution
w = width of elution curve
x = relative concentration in fluid phase
y = relative concentration in solid phase
z = number of separating spaces in disk–type centrifuge
α = selectivity factor
β = intercept of salting–in line and normalized solubility line extended to zero concentration
γ = surface tension at solvent–air interface
δ = Freundlich isotherm exponent
ΔG_{sol} = change in free energy
ε = void fraction in packed column
κ = mass transfer coefficient
Λ = slope of salting–in line
μ = viscosity
ν = velocity
ρ_m = density of fluid medium
ρ_p = density of particle
ρ_s = density of solid phase
σ = standard deviation
σ_i = molal–surface–tension increment
= equivalent clarification area
φ = angle of separating space in disk–type centrifuge
ψ = liquid–solid contact angle
ω = angular velocity
Ω = intrinsic salting–out coefficient

FOREWORD FOR SCOTT WHEELWRIGHT'S BOOK

If one looks at the evolution of great technological innovations, a number of common patterns appear. Two such patterns are the increasing breadth of applications and the increasing economy of using the technology. The evolution of the biotechnology industry has certainly witnessed these trends. Actually, these two trends are probably linked: as the technology becomes more economical it is used more broadly, and as it is used more extensively it becomes more economical.

At the start of the computer age, single computers were the size of large rooms and were used to calculate ballistic trajectories. If one follows the expansion of the computer industry over the last few decades, it is astounding what computers are being used for today, from the ultrasophisticated to the mundane. Airplanes are being built today that cannot be flown by man but must be flown by computers. Medical imaging techniques, such as MRI and CAT scanning, are only possible because of computational power and programming sophistication. Most cars, television sets and microwave ovens have computers or at least microprocessors. Even a $25 wrist watch can have a chip that not only keeps time but is a calculator as well.

While biotechnology is a much younger industry, it already appears to be following the same trends. A decade ago there were no biotech pharmaceuticals and today there are ten. In ten more years, at the turn of the century, there will be many more. More than half of the new drug substances now being submitted for approval to the FDA Center for Biologics Evaluation and Research are produced using either recombinant DNA–derived cells or hybridomas. Even many conventional drug substances are now being discovered and developed using recombinant DNA and monoclonal antibody technology. Receptors and other biological targets such as the HIV polymerase are being purified from cloned sources rather than from naturally–occurring materials.

Cesar Milstein has said that when he and George Koehler discovered monoclonal antibodies in 1976, they did not imagine the myriad uses for these molecules. Fortunately for the rest of us, they published their important work without applying for a patent. Today, monoclonal antibodies are being used as assay reagents in research as well as for clinical diagnostic tests. These tests have become simple and cheap enough to be used in the home, such as the early pregnancy test. Monoclonal antibodies are also being used as "magic bullets"—molecules that can specifically seek out targets in the body to deliver potent drugs for a concentrated local effect. Sometimes, the antibody itself is the drug, as in the case of OKT 3 or individualized monoclonals against B–cell lymphoma. Other times, the antibody is coupled to another effector reagent as in the case of radio-labelled antibodies or antibodies which are coupled to a toxin such as Ricin A chain. In these cases the antibody's purpose is to direct the site of action of the other entity.

Monoclonals are also being used in the purification of many other products, especially recombinant proteins. Progress in this field has been swift and dramatic. Improvements in the culturing of hybridomas, attachment of monoclonals to support media, and elution and regeneration procedures that extend lifetime of this media have reduced the cost of using monoclonals by a factor of about one thousand over the past decade. This improved economy coupled with the unique ability to tailor the specificity of monoclonals puts them in the forefront of our recovery and purification tools.

In the same way that monoclonal antibodies can be used to purify recombinant proteins, recombinant proteins can be used to purify monoclonals. Protein A is now made by recombinant DNA technology. Another microbial protein that binds monoclonals, Protein G, has been mutated to delete the albumin-binding domain. This makes it useful as a capture reagent even in the presence of whole serum. This is an example of how the availability of a new technology can act to accelerate its own development in much the same way that each generation of computers is used to design the next generation.

In this historical context, what is the place of the bioprocess industry within the broader biotech industry? A timely example is available in the development of Activase™.

When I started working on the development of the process for TPA in 1982, the dose was estimated to be one milligram. The capacity and cost per gram targets based on this dose estimate did not seem too difficult to achieve. But very shortly the estimates of dose began to rise and there was a simultaneous shift away from an *E. coli* expression system to a mammalian cell culture process. At the time, these changes made the task appear quite a bit more difficult, but still possible. As experience was gained in human clinical trials, the dose continued to rise until it reached 100 mg! If, in 1982, we had known that this was to be the dose, we probably would have given up. Fortunately, progress on the process was made at about the same pace at which the dose rose. Thus, while the dose was rising by 100-fold, the cost per gram was falling by about 100-fold, keeping the cost per dose about the same. If not for all the progress in molecular biology, fermentation and purification science, and manufacturing technology, tissue plasminogen activator would not be available today. Even more important is the fact that this technology is now being used to accelerate the development of a significant number of new molecules as they work their way through development and regulatory approval.

I suspect that the ultimate size of the biotech industry will, to a significant extent, be determined by the success of bioprocess scientists and engineers in making bioproducts at low cost, large scale, and high quality. This will permit the broadest use of these products. There are several projects now under way that represent the current forefront of this effort. One of these, bovine growth hormone, is nearing approval. The market need may be as great as 100,000 kg per year and the production cost has to be 1 to 2 orders of magnitude lower than that for current recombinant human pharmaceutical proteins. Still to be mastered are the challenges presented by recombinant albumin, recombinant hemoglobin and certain monoclonal antibodies given in multigram doses. The production of these proteins will require another order of magnitude improvement in economy and probably at least a similar improvement in purity. While it may be hard to imagine today exactly how this will be done, I am optimistic that great progress can be made.

Scott Wheelwright's book is an unusual accomplishment. It is rare today for an individual scientist to write a book of such breadth and clarity. In addition to covering the fundamentals of the properties of proteins and the unit operations of protein purification, he also addresses the important related issues of process design, integration with upstream operations, cost, and timing to market. I believe that this book can serve as a foundation for someone trying to enter this field or as an enjoyable review for current practitioners. The figures should make it a useful reference source. Hopefully, it will help in the training of more people capable of contributing to the advancement of bioprocess technology. It is my expectation that this in turn will result in the continuing expansion of the whole biotechnology industry.

Stuart E. Builder
So. San Francisco
January 7, 1991

CHAPTER 1

PROCESS DESIGN AND COMPETITIVE ADVANTAGE

The design of a process for the large–scale purification of a protein is one of several complex activities that must be successfully completed to bring a product to market. The complexity is due in large part to the many alternatives from which the designer must select the appropriate recovery steps; but the design is also complicated by its impact on the competitive advantage of the corporation. This book is intended to help the process designer optimize the design so as to provide competitive advantage for the strategy of the corporation: The process is to be designed to give the company the advantage it seeks in the marketplace.

There are three main sources of competitive advantage in the manufacture and sale of high–value protein products: first to market, high quality (in terms of purity, activity, dependability, or flexibility), and low cost. These alternatives are illustrated in Figure 1.1. Each of these offers advantages that can be realized only at the expense of the others. To decide which process steps are appropriate, the company must define the advantage sought. The optimal process is that which gives the best performance with respect to the desired advantage.

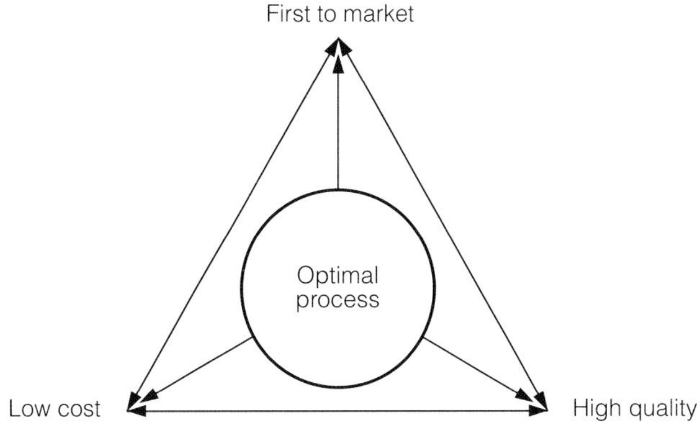

Figure 1.1. The optimal process must be defined within the context of competing goals or sources of competitive advantage.

1.1 SOURCES OF COMPETITIVE ADVANTAGE

Competitive advantage may arise from many sources. Anything that provides an edge over competitors or serves to differentiate the product from similar products is a source of advantage. As described by Porter (1985), in the world at large the opportunities for creating competitive advantage are almost limitless. Within the field of high–value proteins the three categories already mentioned cover most of the common sources of advantage.

Perusal of business journals that cover the industry shows that the company that is the first to market a new protein enjoys a substantial advantage over the competition, particularly if being first to market includes being first with patent protection. Large sums are spent in drug research, for example, to be first with blockbuster money makers. The company that finds itself in second place may net only one–tenth the sales of the company that got there first. With this difference, it obviously makes little sense to spend years developing the lowest cost process if long–term development means losing the race. On the other hand, if the product is not unique or if it can exist in many forms, then a greater advantage may be obtained by differentiation of quality (for example, activity or dependability) than by being first to market.

Many products must share the market with competitors for various reasons, such as insufficient patent protection to keep all competition off the market. In such cases being first to market may not be enough to ensure competitive advantage, and product differentiation may be obtained by varying the characteristics of the product. A protein that is more pure, is more active, has greater lot–to–lot consistency, or is improved by some other measure of performance may outshine the competition. Similarly, a company that can configure its product in more alternatives to suit individual customers shows greater flexibility and may gain an advantage.

As a product acquires competitors, price becomes a greater issue with customers, and lower–cost products gain an advantage. Of course, the lower the cost is from the beginning, the higher the margin and the greater the profit. Thus the company whose process designer has initially developed a lower cost process prevails over the company that must redesign to lower its cost. Regulatory restrictions impose such a high cost on process change that good original design provides a competitive edge.

Once the source of competitive advantage has been identified, it must be maintained as the criterion for evaluation of process design because it defines the goal. The optimal process is the process that minimizes time to market, maximizes a differentiating characteristic, or minimizes cost within the constraints of product specifications and available resources.

1.2 PRODUCT SPECIFICATIONS

With this definition of the goal in mind, we can review the product specifications. The purity required of the product depends on its intended use, as in Figure 1.2, which shows the increase in purity expected in products as a function of dosage. As the dosage increases, so does the purity requirement. Agricultural products and specialty chemicals require only 90 to 95 per cent purity, as do in vitro diagnostics. In vivo diagnostics and other parenterals, such as vaccines, that require only a few injections, do not need so high a purity as do drugs that are administered in large quantities.

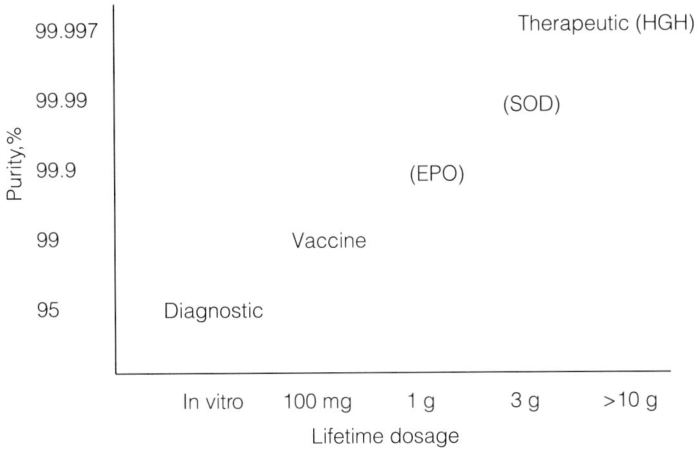

Figure 1.2. Required purity as a function of dosage for selected examples of biotechnology products.

The level of purity required also varies with the nature of the contaminants: Endotoxins, nucleic acid, and anaphylactic agents must be reduced to very low levels, whereas precursors to, or fragments of, the product may be allowable at higher levels.

1.3 FIRST TO MARKET

Being first to market requires short process–development times. In a regulated industry (such as pharmaceuticals) it also means that approval must be granted first. A sloppy process that does not provide consistent product may result in delayed product licensing. This delay can be longer than it would be in the case of a better process. Of course, selecting early license application or more complete process design is a judgment that must be made on the basis of experience and is part of what makes process design complex.

In a regulated industry changes in manufacturing methods tend to be expensive, as they often require repeated studies of safety and efficacy as well as delays resulting from seeking approval for the changes. Builder et al. (1988) estimate, for a licensed product, a minimum of two years from the time of the conception of a process change until the point when a license amendment may be granted and a cost for the entire effort of millions of dollars. Thus designs may be difficult to alter, and rushing in with an incomplete or ill–defined process may cause considerable difficulties once the product is on the market.

1.4 QUALITY

Quality has almost as many different definitions as there are people who define it. A practical definition in the production of biologicals is that the higher the quality, the less the loss incurred by the customer in using the product. As stated by Taguchi (1986), the magnitude of quality is inversely proportional to the loss caused by the use of that product. In biologicals this frequently means that consistency between lots is more valuable than increased purity; a customer incurs less loss when each batch of the product is equivalent (say, 95 per cent pure) than when the purity (and impurities, side effects, and the like) varies (say, from 92 to 98 per cent). As Taguchi writes, "an article with good quality performs its intended functions without variability, and causes little loss through harmful side effects, including the cost of using it. If cost control is concerned with reducing the various losses that may occur before the product is shipped, quality control is concerned with reducing the losses that it may cause to society after it is shipped." Anything that can be done to establish the quality of a product when it is still in the design stage is many times less expensive than trying to fix problems in the manufacturing stage.

The quality of the product is dependent upon the process by which it is made. Quality cannot be inspected into a product but must be designed into it via the process. Two important factors enter into design for quality: design for manufacturing and robust design.

The unit operations available for protein purification vary in many ways. Some are limited to small–scale operation, others to large–scale operation; some have few operating variables, others many; some are simple to control, others difficult. Design for manufacturing means that the limitations of large–scale manufacturing are considered before the process is implemented in the factory. Of particular significance in the design for manufacturing is the number and complexity of operations. Simple operations that can be precisely controlled are less expensive in terms of both product cost and quality than are difficult and complex procedures.

Variation is inherent in biological processes. Organisms have different metabolic pathways that result in the formation of different products under varying conditions. Because variations are the rule in producing biological products, the design of a process to manufacture product with consistent quality poses special challenges. The design must allow for a certain level of variability (noise) in the incoming materials and yet produce a uniform product. Various unit operations may be termed "robust" in that they are relatively insensitive to input variations (for example, affinity and ion exchange chromatography) whereas other operations exhibit appreciable change in the output because of input variations (for example, gel filtration chromatography). The process that consists of robust steps is likely to produce consistent product under naturally varying conditions.

Design for manufacturing and incorporation of robust processes also have an advantage in technology transfer. At some point the process must be transferred from the development laboratory to the factory. At that time the process will move from one environment to another. The steps and conditions of the process can be transferred, but the work environment of the laboratory will not be transferred. Significant difficulties may be encountered in the transfer if the differences in the environments are not acknowledged and if all critical steps in the operation are not clearly specified with allowable limits. Frequently the process is not fully understood until after the transfer. The incorporation of robust processes and the concept of design for manufacturing simplify and shorten the time needed for complete technology transfer.

1.5 PROCESS COSTS

The costs of producing a protein product may be grouped into four broad classes: expenses, labor, capital equipment, and facilities. As shown in Figure 1.3, the total cost is a function of these four cost categories.

$$\text{Total cost} = f(\text{expenses, labor, capital, facilities}) \qquad (1.1)$$

The process designer affects the costs in each of these areas through the selection of process steps, the efficiency of those steps, the scale of the equipment, and the degree of automation. The skill of the process designer is most apparent in the selection of process steps and the optimization of

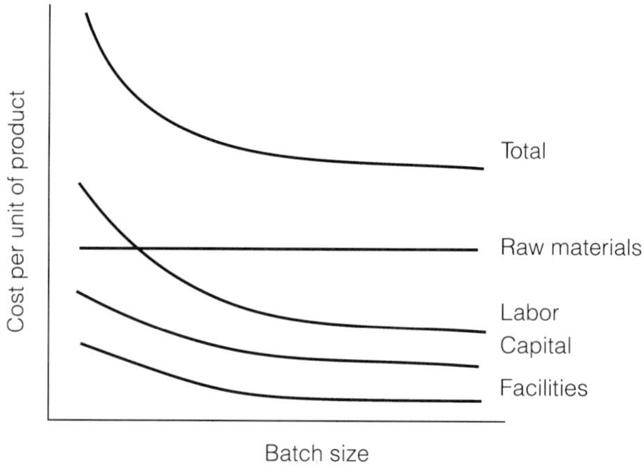

Figure 1.3. The total cost is a function of raw-materials expense, capital equipment, labor, and facilities, each of which varies with the batch size.

those steps for high efficiency (with respect to both product yield and raw–material consumption). Processes with fewer, simpler steps usually cost less to operate, and increased yields reduce the cost of production.

The scale of the equipment, the degree of automation, the batch size, and the amount of labor are all competing factors. If the process can be easily automated, smaller equipment can be used with smaller batch sizes, rapid turnaround, and minimal labor. If the process is not easily automated, larger batch sizes reduce the labor cost (which includes not only production operators but also quality–control release testing for each batch) but increase the capital cost. For marginal increases in batch volume, the cost of labor may be constant while the cost of capital equipment increases with an exponential power less than 1. An increase in the number of batches directly increases the labor cost without increasing the equipment cost. The tradeoff must be decided in light of the ultimate scale and cost of automating a process. Difficult processes of small volume can be produced by a skilled operator for less than the cost of automation; large volume simple processes can be automated for less than the cost of highly skilled labor.

Another consideration at this point is the life cycle of the product. The majority of products on the market today will not be on the market ten or twenty years from now. Some products, such as in vitro diagnostics, have a market life of less than five years. The life expectancy of the product should also be considered when choosing the equipment: Long–lasting equipment should be flexible in application so that it can be used with future products.

The facility cost is the least sensitive to variations in the process. A clean work environment suitable for the product being produced with sufficient floor space to hold the necessary equipment must be furnished. The choice of the particular equipment to be installed has a minimal impact on the total facilities cost, largely because the number of alternative types of equipment is limited, and utility and space requirements for different designs tend to be similar. A plant dedicated to the production of a single product will give the lowest cost per unit of production; multiple products and pilot plants for production of clinical–trial material require increased flexibility and more equipment, which is in use only part of the time, leading to higher costs as measured per unit of production.

It is difficult to accurately estimate the cost for a downstream process before the process is fully developed and yields have been measured. However, some processes are inherently more expensive than others, owing to expensive raw materials (for example, affinity chromatography), expensive equipment (for example, ultracentrifugation), or intense labor (for example, multiple extractions). In spite of the high expense, each of these methods has an application where it is the alternative that gives a higher yield and lower cost overall. In the design of a process for low cost it is the total cost that matters: Incorporation of one expensive but high–yielding step may be less costly than are several less expensive but low–yielding steps.

Most often the large–scale process is developed by making small changes in the method used to generate research material. The research process is used "because it works." A more cost–effective method is to consider each possible alternative thoroughly before deciding on the final process. The designer can do this on paper, listing the factors for and against each option, including approximate costs, before ever trying them in the lab. The most important point to keep in mind is the long term advantage gained by having a less costly process, within the restrictions initially set with regard to purity.

1.6 FURTHER DESIGN CONSIDERATIONS

Further considerations related to the design of a large–scale process that have an impact on competitive advantage include the following:

1. *Documentation*: If others will be running the process, it must be fully documented in such a way that they can produce the same material by following the documented directions.

2. *Reproducibility*: The product of this process must be reproducible in a different environment. The process can be transferred, but the environment in which it was produced, that is, the minor details and variation of

techniques, cannot be transferred. The designer of a process must keep in mind a need to define the points that are not critical and a need to set tolerance limits on all points, wide ones on those points which are not critical and narrow ones on those which are more critical.

3. *Sanitation*: Since pharmaceuticals must be safe and sterile, one must eliminate pyrogenicity, contamination, and infection with other materials and growing organisms; see Schwartzberg (1987). This also includes the incorporation of specific steps for the removal and inactivation of virus associated with production in cell culture.

4. *Virus clearance*: The potential exists that mammalian cell lines may contain endogenous retroviruses that could potentially lead to patient harm. Limitations in the sensitivity of analyses for viruses has led to the incorporation of specific steps (such as heat, extreme pH, detergent addition, and so forth) that are capable of reducing the amount of virus by inactivation or removal. Builder et al. (1988) reported that for TPA, based on a calculation to give a final concentration equivalent to 0.01 viral particles in one million doses, a reduction of 12 orders of magnitude was appropriate. Due to assay limitations they could validate only 4 logs of clearance in a particular step and therefore included three independent mechanisms for virus reduction. Fenno et al. (1988) have described the evaluation of a process step for its efficiency at clearing virus by the addition of a known amount of virus prior to operation of the step followed by measurement of residual viral infectivity after the step.

5. *Safety and containment*: Some materials that are processed are biologically hazardous and require containment. Safety is always a concern in the design of the process, especially in steps such as liquid heating or dangerous–material handling. We must keep in mind that the ultimate scale is one of the criteria for design, and that the large quantities we are dealing with influence various safety aspects.

1.7 OVERVIEW OF BOOK

Once the goal and the optimization strategy have been defined, the technical challenge begins: How do we choose between the alternatives and, beyond that, how do we choose the conditions for each step and optimize recoveries, yields, and purity?

This book seeks to answer these questions. Chapter 2 covers the first question, strategies behind alternative process designs, including physical bases for separations and rules of thumb for selecting the order of various unit operations. Chapters 3 to 5 review aspects of protein chemistry, protein analysis, and fermentation operations that bear on the purification operations. The remaining chapters review the various unit operations that are used in the recovery and purification of proteins at large scale. Emphasis has been placed on those techniques for which commercial equipment is available. The last chapter provides examples of large–scale processes, including data on recoveries and purities, where available.

REFERENCES

Builder, S. E., R. van Reis, N. Paoni, and J. Ogez, "Process development and regulatory approval of tissue–type plasminogen activator," paper presented at 8th International Biotechnology Symposium, Paris (1988).

Fenno, J., J. Luczak, A. Moore, J. Poiley, R. Raineri, and M. Whiteman, "The characterization of transformed cell substrates," *BioPharm* 1(8):36–42 (1988).

Porter, M. E., Competitive Advantage: Creating and Sustaining Superior Performance, Free Press, New York (1985).

Schwartzberg, H. G., "Process and plant design for food and biochemical production," in Applied Protein Chemistry, R. A. Grant (ed.), Applied Science Publications, New York (1987).

Taguchi, G., Introduction to Quality Engineering, Asian Productivity Organization, Tokyo (1986).

FURTHER READING

Hayes, R. H., and S. C. Wheelwright, Restoring Our Competitive Edge: Competing Through Manufacturing, John Wiley & Sons, New York (1984).

Hayes, R. H., and S. C. Wheelwright, "Link manufacturing process and product life cycles," *Harvard Business Review* 79(1):133–140 (1979).

Porter, M. E., Competitive Strategy: Techniques for Analyzing Industries and Competitors, Free Press, NY (1980).

Reisman, H. B., Economic Analysis of Fermentation Processes, CRC Press, Boca Raton, Florida (1989).

CHAPTER 2

STRATEGIES FOR PROCESS DESIGN

As biotechnology moves from the research lab into the marketplace, the need for large–scale production processes becomes obvious to all. Many recombinant proteins are currently available as drugs, for example, insulin, interferon, hepatitis B vaccine, tissue plasminogen activator, and human growth hormone. Many more biotech–produced proteins are expected soon, including growth factors and enzymes. But manufacturing must keep pace with the development of these products and their movement into the marketplace. To manufacture these proteins, one must have a process, and it must be a process that operates on a large scale, producing large quantities economically. The question then is how one designs a large–scale process for protein purification. Or, in other words, is there a rational method for the design of downstream processes?

The answer is yes and no. Yes, in the sense that many examples of successful processes exist, and guidelines in the form of suggestions also exist. But no, in the sense that there is no algorithm by which one can proceed, step by step, to develop a foolproof large–scale design. The number of options for protein purification is so large that one cannot say that there is only one right method or that one best process exists. In addition to the number of options considered, each protein is different, and these differences among proteins preclude the formal development of an algorithm for designing downstream processes. The purpose of this chapter is to present, in the context of historical background, current trends in process design. Alternate approaches to general process design and an approach to the design of downstream processes are described.

2.1 DEFINITIONS AND BASES FOR DESIGN

"Downstream process" may be defined as *a series of steps that, when followed, results in a purified protein product.* The definition of an "optimal" process in economic terms, that is, with respect to purity, cost, and market entry, should not be confused with the optimization of parameters within the unit operations. Process design and scale up may be divided into three distinct steps: selection of unit operations, optimization of operating parameters at the small scale, and increase of equipment size while maintaining the same operating parameters. (The phrase "unit operation" refers to a distinct chemical or physical step such as centrifugation, chromatography, or ultrafiltration.)

Scale up, in the historical sense, has usually referred to the last step, the increase in batch size of an existing process. That is, one starts with a process that has been used to produce material at small scale and then increases the size of each of the unit operations in that process. Thus, historically speaking, the design of processes for large–scale production has not been done.

An important point to consider in the design of a purification process is the market volume. That is, how large the market is and what will be the ultimate scale of production. Smith (1968) claims that the starting point for process design must be the ultimate commercial scale and that process development activities must relate to the scaled down version. By considering the ultimate scale from the start, we may avoid a common trap, that is, the use of unit operations that are difficult or expensive to operate on a large scale. Our goal is not to increase the size of operations performed at the bench scale; our goal is to develop an economical large–scale process. We begin by specifying the ultimate scale, based on market forecasts; this will dictate the volume of our operations. Then we scale down our proposed process for the purpose of feasibility studies and for optimization of each unit operation. Bisio (1985) further states that "the only defensible reason for scale up studies is a reduction in the possibility of making expensive errors in the design or the operation of commercial size equipment." He further suggests that the scale up research should be focused on reducing uncertainties faced by future design engineers and plant operators.

As a sidelight, it is interesting to note that the idea of looking at the ultimate usage when defining the beginning process is not restricted to downstream process design. Hart (1988) describes the application of the concept by the developer of a pest–control service with an unconditional guarantee. Hart explains that the developer "began with the concept of the unconditional guarantee and worked backward, designing his entire organization to support" the final product. In short, he started with the vision and ended with a means to achieve that vision.

In addition to defining the design goals, one must collect the necessary background data for developing the design. Ratafia and Keenan (1986) suggest that items such as solubility data in regard to salt solutions, organic solvents, and pH, size, shape, net charge, and distribution of charge, and density, adsorption isotherms, and detergent binding may all be important chemical and physical data to collect before designing the process. Wheelwright (1987) suggests in addition that the major contaminants, the presence of different phases, and physical characteristics of the product such as thermal stability, isoelectric point, and hydrophobicity are helpful in developing the design process. In addition to chemical and physical data, thermodynamic considerations may need to be taken into account. Kabel (1985) suggests that thermodynamic analysis makes possible calculation of the maximum conversion available in any step as well as guidance in the formation of rate equations and heat effects, where they occur.

2.2 DESIGN METHODS

The most common method for downstream process design is to copy an existing process, that is, to start with an available method that has been shown to produce one product and adapt it for a similar product. There are many purification methods in the literature. Although most of these are for the small scale, some are available for larger–scale processes, and some may be easily adapted to the large scale. The advantage of copying an existing process is that it is the simplest method by far. It requires less development, less expense in terms of time and equipment, and so on, to produce the required product. If the protein for which we are trying to design a process is similar in nature, characteristics, and properties to other proteins for which we have existing processes, then copying or transferring those existing methods may be very straightforward and simple. But every protein is different, and there is no guarantee that this will produce acceptable material. Again, the product we are trying to purify may be unrepresented in the literature; if that is the case, then we must have another method for developing a process design.

The second method, both in terms of most common application and in order of complexity, is called "heuristics," or rules of thumb. Just as the use of the thumb, as a measure of length, results in a crude approximation, heuristics serve as guidelines only for approximate design. Heuristics require the designer to exercise expertise in the selection of processes, while providing a type of distilled collective wisdom to direct him or her along the most productive paths. Resnick (1981) recommends that heuristic reasoning be considered not as a final or strict method but rather as a search for what is provisional or plausible — that is, seeking a solution to a problem by means of plausible guesses, which may be fallible guesses. The purpose of heuristics is to narrow the list of possible separation steps based on general experience. A list of heuristics allows a designer with limited experience or with experience in a limited area to call upon the general experience of others who have worked in the field for the selection of available possibilities in process design.

There is ample precedent in the literature of the chemical industry for the use of heuristics for multicomponent separations. Nadgir and Liu (1983) have classified heuristics for multicomponent separations in terms of their applications, and they separate them into four broad categories: (1) *method heuristics*, or rules for specifying a choice between different unit operations, (2) *design heuristics*, or rules for specifying the order or sequence in which the steps should be performed, (3) *species heuristics*, or rules based on the properties of the components, and (4) *composition heuristics*, or rules related to the influence of feed and product composition on the separation costs. These classifications are summarized in Table 2.1. (A list of heuristics used for chemical–separation sequences is included in Appendix 2.1)

Table 2.1. Classification of heuristics.*

Type	Application	Example
Method heuristics	Specify choice between different unit operations	Describe conditions for preference of ultrafiltration over centrifugation
Design heuristics	Specify order or sequence of process steps	Perform low resolution step prior to high resolution step
Species heuristics	Specify step based on properties of components	Choose separation method based on greatest difference in properties of product and impurities
Composition heuristics	Specify product of feed composition based on separation costs	

* Based on Nadgir and Liu (1983).

A third approach in process design is the use of expert systems. The idea here is to follow the pattern of thought of the expert designer through a series of if–then postulates. Lien et al. (1987) described this as a "mixture of opportunistic qualitative and quantitative analysis mixed with goal and constraint posting, replanning checking, etc." Expert systems are based on models describing the expert; they do not necessarily follow a strict algorithm but rather rely more on qualitative judgments for choosing among the competing alternatives.

The fourth design method is the algorithm, or strict mathematical model. (A model is a mathematical representation of a physical process.) The use of a model allows investigation of the effects of operating condition changes on the process output. The difficulty with using models to describe highly complex systems is that a highly complex model may be required. Denn (1986) suggests, however, that even a simple fundamental model has value in the development of heuristics and empiricisms.

Our fifth choice is some combination of the preceding approaches. Westerberg (1987) defines process synthesis as "the discrete decision making activities of conjecturing (1) which of the many available component parts one should use, and (2) how they should be interconnected to structure the optimal solution to a given design problem. The challenge may be to create a new process (grass roots design) or to improve an existing one (retrofit design)." Thus synthesis may consist of following an existing example, modifying it by means of heuristics, following an expert system, following a modified expert system that relies mainly on heuristics, or having an algorithm for part of the design and using these techniques to combine the unit operation in a stepwise fashion. Thompson and King (1972) use a synthetic combination of heuristics to describe multicomponent–separation process design and comment that "process synthesis is regarded for the most part as intuitive art." Seader and Westerberg (1977) have outlined four basic ideas or steps in evolutionary synthesis, namely, (1) an initial flowsheet, (2) rules or heuristics for making systematic or small changes to the flowsheet to create a neighboring flowsheet, (3) a strategy for applying these rules, and (4) a means of comparing the original flowsheet to any of its neighbors.

It is important to note that, although we have listed four distinct methods of process design and a fifth, which is a combination of these methods, there is no strict division between the approaches, and overlap occurs. That is, they all rely to some extent on a model. As we progress from the simple method of copying an existing process, through heuristics and expert systems, to the complicated algorithm, the model becomes more stringent and more complex.

2.3 EVOLUTIONARY SYNTHESIS

In practice, it is safe to say, the design of a large–scale protein–purification process is an evolutionary method that relies, where possible, on an existing process and applies a set of general rules, with some degree of expertise, to the development of a series of steps to effect the purification. To begin our design, we must have a starting place, and that starting place for purification is the ending place for the fermentation or synthesis step. Figure 2.1 describes a general downstream–process flowchart. The customary definition of "downstream" includes everything after product synthesis. Thus downstream processing begins with the recovery operation. The various stages of downstream purification have been described differently by different authors, but they generally fall into the categories depicted in Figure 2.1, that is, recovery, low–resolution purification, and high–resolution purification. Belter et al. (1988) refer to these as (1) removal of insoluble material, (2) isolation of products, (3) purification, and (4) polishing.

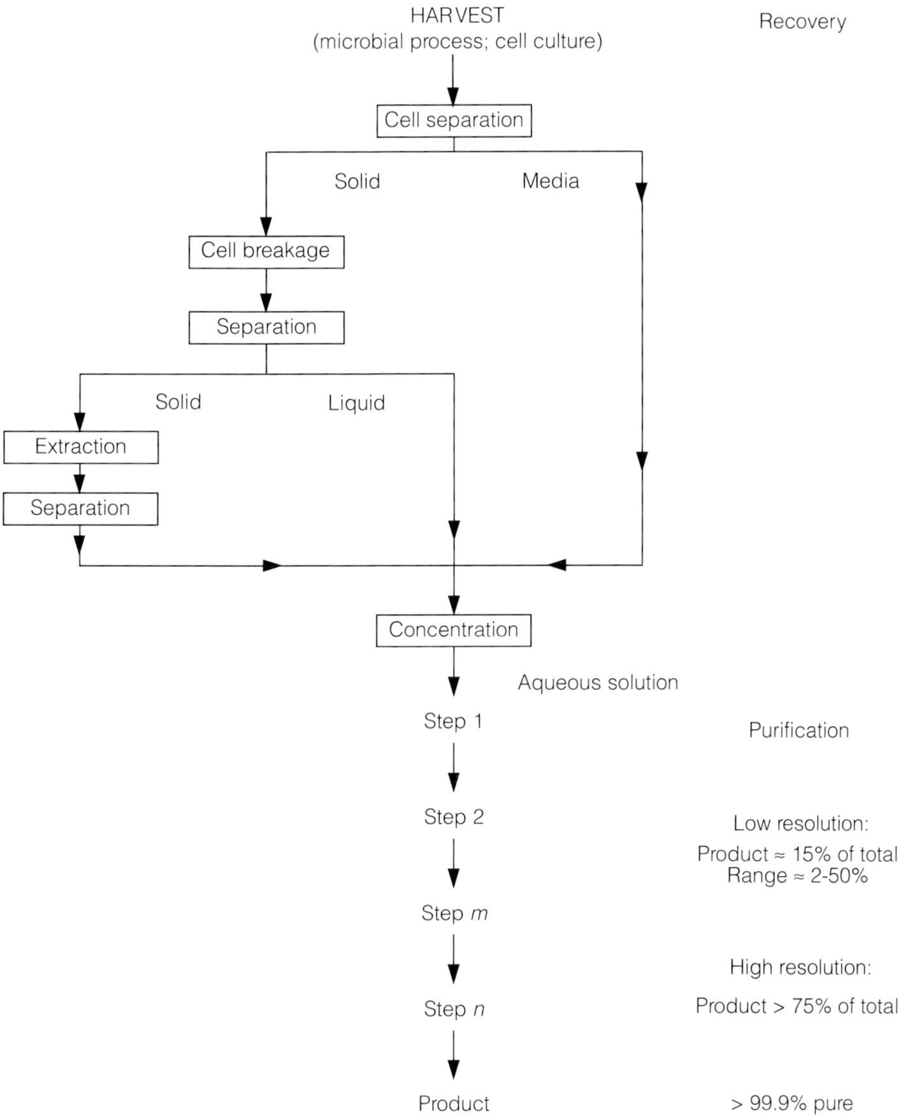

Figure 2.1. Flowchart of downstream processes, illustrating steps in recovery and steps in purification.

Similarly, Lightfoot et al. (1987) refer to three classifications, namely, (1) concentration (increased concentration of the valuable species), (2) fractionation (separation of two species close in value), and (3) purification (the removal of impurities at low concentrations).

As shown in Figure 2.1, the recovery stage starts with the spent medium and ends with an aqueous solution containing the protein product. In the case of tissue culture or if the product is secreted as a soluble species, the recovery step consists solely of a concentration step to remove excess liquid. However, many product proteins are insoluble and are located inside the cell mass, and therefore an extensive recovery procedure is required to break the cells, separate out the debris, and recover the product protein, either from the liquid inside the cell or from the cellular debris by extraction with detergents or otherwise. Once a soluble concentrate is obtained, purification consists of steps that increasingly remove impurities and increase the relative concentration or purity of the product. The choice of purification steps is the major challenge in the design of purification processes.

With regard to recovery, there are basically few choices among the different options. Either the material is secreted or it is not. If it is secreted, then a concentration step is all that is necessary; the choice there may be ultrafiltration or adsorption. If it is not secreted, then the packed cells must be broken, and once they are broken, then we again have a solid–liquid separation step, which may be ultrafiltration or centrifugation. If the product is in the liquid phase, then concentration may again be necessary, either by ultrafiltration or adsorption. If it is in the solid phase, then an extraction is necessary with detergents or solubilizing agents until a concentrated aqueous solution is obtained. Therefore in the recovery the choice is mainly in the types of equipment for the different unit operations rather than between multiple unit operations. The division of the steps into recovery and purification is arbitrary, and in particular instances a different classification may be preferred. The material becomes more pure as it moves from one step to the next, the total mass of protein decreases, and the relative amount of product increases. And as the product becomes more pure, the separation of impurities becomes more difficult. One of the hardest separation problems in recombinant products is the issue of microheterogeneity, or the separation of active and inactive or fragmented isomers.

2.4 SELECTION OF PURIFICATION STEPS

Here is the heart of the problem. How does one choose among unit operations for purification? That is, how does one specify those steps which should be used and the order in which one should use them? The first step is to generate alternatives. As Resnick (1981) puts it, the designer "must generate as many alternative solutions as possible and then analyze and screen them to eliminate poorer ones while generating alternative solutions. Thinking should not be restricted and imagination should be given a loose rein." One way to generate alternatives is to gather together the people involved and have a brainstorming session to list as many as possible.

Another way to generate alternatives is the use of morphological analysis. Resnick describes morphological analysis as

> the rigorous examination and evaluation of all possible alternatives for each part of a problem. It aims to produce the optimum solution by virtue of having considered and evaluated every possibility. The analysis provides a logical structure that can replace random thinking, is comprehensive, helps to ensure that all alternative solutions are being considered, and presents the possibility of new product syntheses resulting from combinations of alternatives that had not been previously considered.

A thorough description of the morphological approach has been presented by Zwicky (1967). His definition of morphological analysis is "nothing should be accepted as impossible unless it is clearly proved to be impossible." A method Zwicky recommends is the construction of a multidimensional matrix (examples for ultrafiltration and chromatography are given in Chapters 9 and 10). The five steps of this process are (1) the problem must be exactly defined, (2) all the parameters that influence the decision or the solution must be characterized and listed, (3) the multidimensional matrix, which contains all of the possible solutions of the problem, is constructed, (4) all the solutions are contained in this matrix, and they must be analyzed and evaluated with respect to the problem and goal, and (5) the best solutions are selected and tested by implementation.

Once the possible alternatives have been listed, then heuristics may be used to discriminate between the alternatives. Heuristics for protein–purification design may be divided into the categories described by Nadgir and Liu (1983) as listed in Table 2.1, that is, method heuristics, design heuristics, and species heuristics. (Composition heuristics rarely apply in biotechnology, where feed composition is not subject to discretionary change and products are singly purified, not recovered as mixtures.)

The majority of the heuristics that have appeared in the literature are of the second type, that is, design heuristics, which recommend an order in which steps should take place: Table 2.2 compares the heuristics of several authors. It is interesting that these are all so similar. Either everyone copies from everyone else or everyone agrees; perhaps these are good guidelines. Basically they suggest that the first step should be of a nonspecific type, one that has a low resolution: unit operations such as precipitation, crude adsorption, and so on. This should be followed by high–resolution steps: unit operations that have a higher selectivity, such as ion exchange chromatography or affinity chromatography. This in turn should be followed by a step such as gel filtration.

Table 2.2. Comparison of design heuristics: recommended order of steps.

Reference	1. Choose first step based on:	2. Choose next step based on:	3. Choose last step based on:
Anonymous (Pharmacia) (1986, 1983, 1980)	Differential solubility (crude adsorption)		Selectivity (high-resolution chromatography)
Belter et al. (1988)	Nonspecific separation	High selectivity	Polishing (crystallization)
Bonnerjea et al. (1986)	Precipitation	Ion exchange, affinity chromatography	Gel filtration
Ostlund (1986)	Intermediate purification (ion exchange, adsorption, affinity)	Further purification (chromatography)	Gel filtration
Scawen et al. (1980)	Initial (concentration)	Affinity, ion exchange	Final (concentration, formulation)
Werner and Berthold (1988)	Precipitation		Gel filtration
Wheelwright (1987)	Largest volume or mass removed	Avoid intermediate handling steps (dialysis, concentration)	Most expensive (on a per-unit-of-product basis)

These heuristics represent the views of various authors regarding the preferable order of steps for protein–purification process. In the case of Bonnerjea et al. (1986) they reflect a review of a hundred examples from the literature. From an economic point of view these rules follow a reasonable logic. Larger volumes cost more than smaller volumes, and low–resolution steps tend to cost less than high–resolution steps; so the initial steps, where the volume is usually larger, should use the less expensive lower–resolution methods, whereas later steps, where the volume and amount of impurities are reduced, may economically use the more expensive high–resolution steps.

The transition from low–resolution steps to high–resolution steps, which culminates in gel filtration as the last step, may appear to be incongruous: gel filtration is certainly not the highest–resolution step available. It is, however, very expensive in terms of cost per unit of protein chromatographed, and it certainly shows higher resolution when the sample contains only a few species than it does when the sample contains many species, and therefore it logically appears near the end of the process.

In addition to the above design heuristics, Wheelwright (1987) describes two species heuristics. Rather than suggesting a specific order for steps, these offer guidelines for choosing between alternative unit operations at various process stages. These rules are, (1) exploit the physical characteristic or separation basis (for example, surface charge, hydrophobicity, electromobility, and so on) with the *greatest* difference between product and impurities, and (2) exploit *different* separation bases in successive steps.

Table 2.3 lists various separation methods available for large–scale purification classified on the basis of the underlying mechanism of each, along with the chapter number in which a description may be found. Descriptions of the unit operations for recovery and purification are organized by separation basis rather than selectivity or some other grouping. A given unit operation may be used in various ways with widely different selectivity; for example, ion exchange may be used for crude adsorption in a batch process and for highly specific separation in a column–chromatography process.

2.5 PROCESS SYNTHESIS

With our heuristics in hand, how do we tie the process together? One way is by a flowchart to describe the different steps and the stages of the design process. Figure 2.2 is a combination of the flowcharts suggested by Kaufman (1988), Wheelwright (1987), and Seader (1985). By proceeding stepwise through this flowchart, we cover the major points of design: We define the problem, list the specifications, consider alternatives, rely on heuristics for guidance, include design considerations, and meld it all into an economical process. Another method, suggested by Seader and Westerberg (1977), is to utilize a set of heuristics by which one applies heuristics, that is, a set of methods heuristics for application of design and species heuristics.

As described by Taguchi (1986), the design of a process to manufacture a product consists of three stages: system design, operation design, and parameter design.

Table 2.3. Classification of separation processes by underlying physical bases.*

	Solubility	Sedimentation Rate	Density	Size (MW)	Ionic Charge	Hydro-phobicity	Specific binding	Application
Extraction Processes	extraction 8							proteins
	fractional precipitation 8							proteins
Membrane Processes				filtration 9				particles proteins
Centrifugal Processes		non-equilibrium settling 7						particle removal
			isopycnic banding 7					proteins viruses
Electrokinetic Processes				molecular sieve EP 4	isoelectric focusing 4			proteins
					continuous flow EP 4			proteins
Chromatographic Processes				gel filtration 14	ion exchange 11	reverse phase 12	affinity 13	proteins
						hydrophobic interaction 12		proteins

* Numbers are those of chapters, for further information.

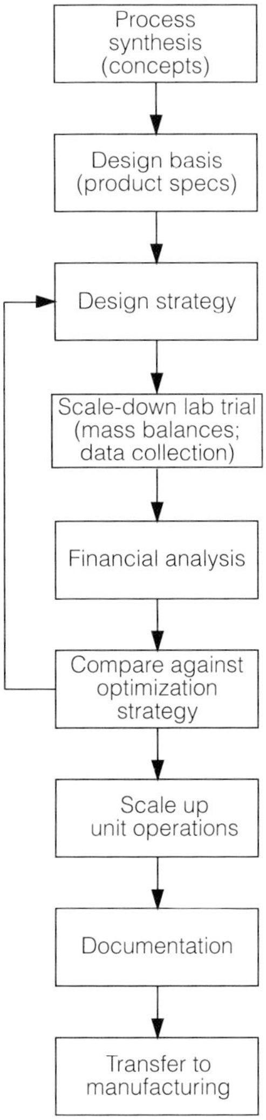

Figure 2.2. Process synthesis flowchart, after Seader (1985).

System design refers to the overall process. Suppose we were making a hormone protein secreted by an animal cell. The overall process might consist of *culture* of the cells, *concentration* of the media, and *purification* of the protein. This is an outline of the necessary steps in the process without any specification of equipment or even method. Based on the product specifications, a general system design describes the *major* sections of the process. Thorough investigation of the alternatives at this stage, prior to definition of the specific operations, can help identify options that might otherwise be overlooked. This initial stage is often neglected by designers or subtly suppressed by management bias and therefore not fully investigated. However, the specification of the system design may have a greater influence on the production cost than the subsequent stages have and therefore deserves primary attention at every level in the organization.

The second stage, operation design, consists of defining the individual steps within the system. Having settled on an animal–cell model, we must choose whether it is grown in suspension tanks, hollow fibers, or by some other method. Do we concentrate by ultrafiltration or adsorption? What type of recovery steps, such as chromatography, do we use? The second stage is the most important when designing a product with low variability and high quality. Here is where we can minimize external and internal noise effects, such as media variations, that alter the nature and amount of impurities. Prudent selection of steps at stage 2 will lessen the variations in product quality resulting from variations in input materials and other sources. It is usually much easier (and cheaper) to reduce product variability at this design stage through adjustment and control of the specific process operating conditions than it is at the next stage. For example, if we expect that variation in media components will vary the number or nature of the impurities, we may wish to use a method such as selective adsorption over, say, size fractionation. Then, when the variability in the crude material appears, we do not have to redevelop the process.

The third stage is that in which we set the actual running parameters. For an ion exchange step, we specify the resin, buffer, and pH. For a cell–culture step, we specify the media components.

All three stages are iterative and overlap to some extent. We cannot determine the overall system without some data on the technical feasibility of the alternatives. We cannopt choose between ion exchange and ultrafiltration without information on adsorption and flux for our product.

Throughout the design process we need to remember that quality is designed in, not added on. By anticipating variability and its causes, we can produce a design that assures the quality of our product.

The development of a downstream process for large–scale production requires the consideration of many competing needs. The synthesis of a process that includes these restrictions is neither simple nor trivial but can be made easier through the application of heuristics, those nuggets of collective wisdom which guide the design process. While heuristics for the downstream processing of chemicals are fairly well developed, heuristics for protein purification are still incomplete; and most large–scale design relies on existing examples and small–scale processes. And remember, heuristics by definition have many exceptions. If they didn't have exceptions, they'd be algorithms.

REFERENCES

Anonymous, "Protein purification step by step," *Separation News* 7, Pharmacia, Uppsala (1980).

Anonymous, Scale up to process chromatography, Pharmacia, Uppsala (1983).

Anonymous, "A strategy for protein purification," *Separation News* 13(6) Pharmacia, Uppsala (1986).

Belter, P. A., E. L. Cussler, and W. S. Hu, Bioseparations: Downstream Processing for Biotechnology, John Wiley & Sons, New York (1988).

Bisio, A., "Introduction to scaleup," in Scaleup of Chemical Processes, A. Bisio and R. L. Kabel (eds.), John Wiley & Sons, New York (1985).

Bonnerjea, J., S. Oh, M. Hoare, and P. Dunnill, "Protein purification: the right step at the right time," *Bio/Technology* 4:954–958 (1986).

Denn, M. M., Process Modeling, Longman, New York, p. 320 (1986).

Hart, C. W. L., "The power of unconditional service guarantees," *Harvard Business Review* 66(4):54–55 (1988).

Kabel, R. L., "Reaction Kinetics," in Scaleup of Chemical Processes, A. Bisio and R. L. Kabel (eds.), John Wiley & Sons, New York (1985).

Kaufman, N., "Scale up in the 1980's," *CHEMTECH* 18:297–299 (1988).

Lien, K., G. Suzuki, and A. Westerberg, "The role of expert systems technology in design," *Chemical Engineering Science* 42:1049–1071 (1987).

Lightfoot, E. N., S. J. Gibbs, M. C. M. Cockrem, and A. M. Athalye, "Scaling up protein purification," in Protein Purification: Micro to Macro, R. Burgess (ed.), Alan R. Liss, New York (1987).

Liu, Y.A., "Process synthesis: some simple and practical developments," in Recent Developments in Chemical Process and Plant Design, Y. A. Liu, H. A. McGee, Jr., and W. R. Epperly (eds.), John Wiley & Sons, New York (1987).

Nadgir, V. M., and Y. A. Liu, "Studies in chemical process design and synthesis," *AIChE Journal* 29:926–934 (1983).

Ostlund, C., "Large–scale purification of monoclonal antibodies," *Trends in Biotechnology* 4:288–293 (1986).

Ratafia, M., and J. Keenan, "Protein separation," *American Biotechnology Laboratory* Nov/Dec:40–47 (1986).

Resnick, W., Process Analysis and Design for Chemical Engineers, McGraw–Hill, New York (1981).

Scawen, M. D., A. Atkinson, and J. Darbyshire, "Large–scale enzyme purification," in Applied Protein Chemistry, R. A. Grant (ed.), Applied Science Publications, New York (1980).

Seader, J. D., and A. W. Westerberg, "A combined heuristic and evolutionary strategy for synthesis of simple separation sequences," *AIChE Journal* 23:951–954 (1977).

Seader, J. D., "Computer Modeling of Chemical Processes," *AIChE Monograph Series* 81:8–9 (1985).

Smith, J. M., "Scale–down to research," *Chemical Engineering Progress* 64(8):78–82 (1968).

Taguchi, G., Introduction to Quality Engineering, Asian Productivity Organization, Tokyo (1986).

Thompson, R. W., and C. J. King, "Systematic synthesis of separation schemes," *AIChE Journal* 18:941–948 (1972).

Werner, R.G., and W. Berthold, "Purification of proteins produced by biotechnological process," *Arzneimittel Forschung* 38:422–428 (1988).

Westerberg, A. W., "Process synthesis: a morphological view," in Recent Developments in Chemical Process and Plant Design, Y. A. Liu, H. A. McGee, Jr., and W. R. Epperly (eds.), John Wiley & Sons, New York (1987).

Wheelwright, S. M., "Designing downstream processes for largescale protein purification," *Bio/Technology* 5:789, 791–793 (1987).

Zwicky, F., "The morphological approach to discovery, invention, research and construction," in New Methods of Thought and Procedure, F. Zwicky (ed.), Springer Verlag, New York (1967).

FURTHER READING

Douglas, J. M., Conceptual Design of Chemical Processes, McGraw–Hill, New York (1988).

Downstream Processes: Equipment and Techniques, A. Mizrahi (ed.), Advances in Biotechnological Processes, Vol. 8, Alan R. Liss, New York (1988).

Schwartzberg, H. G., "Process and plant design for food and biochemical production," in Recent Developments in Chemical Process and Plant Design, Y. A. Liu, H. A. McGee, Jr., and W. R. Epperly (ed.), John Wiley & Sons, New York (1987).

Wheelwright, S. M., "The design of downstream processes for largescale protein purification," *Journal of Biotechnology* 11:89–102 (1989).

Appendix 2 List of heuristics for multicomponent separation sequences.*

1. Method heuristics
   a. Favor ordinary distillation.
   b. Avoid vacuum and refrigeration.
   c. Remove mass separating agent first.

2. Design heuristics
   a. Favor smallest product set.
   b. Use cheapest separator first.
   c. Favor direct sequence.
   d. Remove valuable or desired product last.

3. Species heuristics
   a. Remove corrosive and hazardous components first.
   b. Perform high-recovery separation last.
   c. Perform difficult separations last.
   d. Perform easy separations first.

4. Composition heuristics
   a. Remove most plentiful component first.
   b. Favor 50/50 split.

* From Liu (1987).

CHAPTER 3

PROTEIN CHARACTERISTICS

As a group, proteins have many properties in common that allow their separation from other materials. Individually, proteins have many divergent properties that allow their separation from each other. This chapter describes the nature of proteins, their structure, their interactions, and those characteristics which set them apart from other molecules. To develop a purification method for proteins, we must understand the nature of this class of materials. Proteins are complex molecules; the numerous varieties in which they exist provide many different means for separation and purification. Their similarities allow for general classification and understanding.

3.1 GENERAL BEHAVIOR

Proteins consist of series of amino acids linked together through peptide bonds. All amino acids consist of a central carbon atom linked to an amine group, a carboxyl group, a hydrogen atom, and a subgroup designated R. The amine group and the carboxyl group hooked to the central carbon atom in the alpha position are the basis for the peptide bond. The peptide bond is a covalent linkage between the carbon of the carboxyl group and the nitrogen of the amine group. Therefore a protein has two distinct ends: One is the amino terminal residue, that is, the end having the amino group free (the N terminal); and the other end is the carboxyl terminal residue, the end having the carboxyl group free (the C terminal). Each of the amino acids that make up a protein is called an "amino acid residue," and therefore a peptide can be broken down into the various subunit residues.

Amino acids are classified by their side chain, that is, the R group. It is the variation in the R groups and their attachment to the protein in various orders and configurations that give different proteins their sundry characteristics. Small proteins may have only 50 or 100 amino acid residues, whereas very large proteins may have tens of thousands of amino acid residues. The size of a protein is also measured in terms of its molecular weight. The unit of molecular weight is the dalton, and typical proteins range from 1,000 D (1 kD) to 100,000 D (100 kD) or larger. For globular proteins, 100 kD is approximately 6.5 nm (65 Å) in diameter.

The characteristics imparted to the protein by the amino acid side chains serve as bases for purification processes, and include solubility, charge, hydrophobicity, and intermolecular bonding. Amino acids can be classified by their side chains, as in Table 3.1. The first group classification of amino acid side chains contains the aliphatic and aromatic side chains that tend to increase the hydrophobicity of the protein molecule and make it less soluble in water. The next group comprises the basic side chains, which have an amino group that carries a charge and causes the side group to act as a weak base. A third group includes the acidic side chains, which have a carboxyl group attached and act as weak acids. The fourth group is the sulfur–containing, or thiol, side chains that allow intermolecular bonding. The secondary amine and aliphatic hydroxyl groups take only a minor part in separations, having some influence on solubility.

Table 3.1. Amino acid side chain groupings and separation bases.

Side chain	Amino acids	Separation basis
Aliphatic	Glycine, alanine, valine, leucine, isoleucine	Hydrophobicity
Aromatic	Phenylalanine, tyrosine, tryptophan	Hydrophobicity
Basic	Lysine, arginine, histidine	Positive charge
Acidic	Aspartate, glutamine	Negative charge
Thiol	Cysteine, methionine	Solubility
Secondary amine	Alanine, proline	Solubility
Aliphatic hydroxyl	Serine, threonine	Solubility

The solubility of various proteins differs considerably. Some proteins are very soluble in aqueous solutions. This greatly simplifies their purification. Others are rather insoluble and require special treatment to bring them into solution and separate them from other insoluble impurities. Between these extremes exists a wide range in levels of solubility. The solubility of a protein is also influenced by the presence or absence of sugar groups attached to the protein. Bacteria do not glycosylate, or add sugar chains, to proteins, but yeast cells do for proteins that are secreted outside the cell, as do animal and other cell types for intracellular proteins. These sugar residues tend to be hydrophilic and increase the solubility of the protein molecule. The number and position of the sugar groups depend not only on the host cell line but also on growth conditions. Changing from roller bottles to stirred tanks may alter the glycosylation pattern. Glycosylated molecules vary in the number of sugar residues attached, leading to smearing of the peak on an ion exchange column if the sugars are charged and play a role in binding.

The acidic and basic side chains determine the overall charge on the molecule. That is, as the pH in which a molecule finds itself varies, the acidic and basic side chains will vary their charge.

At low pH the basic side chains will be ionized and contain a positive charge. At high pH the acidic side chains will be ionized and contain a negative charge. Figure 3.1 is a photograph of a computer–generated model of a protein molecule (flavodoxin), with all atoms shown in gray except for oxygen atoms, which are shown in white. While not every oxygen atom will be ionized, at high pH many will bear a negative charge. This picture illustrates the non–uniform distribution of charge over the outside surface of the molecule. Note the tryptophan just left of center.

The point at which the net charge is zero (that is, the number of positive and negative charge groups are equal), is termed the "isoelectric point" (pI). This point is important in protein chemistry for several reasons, the first

Figure 3.1. Model of protein molecule with oxygen molecules (ostensible negative charges) shown in white. (Courtesy of C. Marlowe, Protos Corporation.)

being that the solubility of a protein varies inversely with how close it is to the isoelectric point. Near the isoelectric point proteins tend to be less soluble; farther from the isoelectric point proteins tend to be more soluble. In aqueous solution charge groups interact with polar water molecules and stabilize the protein, which is intrinsically hydrophobic. A low number of charge groups and a high number of aliphatic or aromatic side chains characterize a protein that is less soluble in water. As one moves farther away from the isoelectric point, the number of ionized groups increases and the solubility tends to increase. The isoelectric point then represents the pH at which the minimum number of groups are ionized and where the molecule as a whole is electrically neutral. The isoelectric point can be measured by isoelectric focusing and is a very useful property to know when developing a purification method. According to Scopes (1987), the pI of most plant and bacteria proteins is around 4.5 to 5, while the pI of most animal proteins is around 5.5 to 6.

Proteins are polymers and, like many polymers, tend to be viscous at high concentrations. The viscosity is nonnewtonian and frequently thixotropic, that is, sheer thinning: As the solution is stirred, the viscosity decreases. Proteins are also sheer sensitive; they are long chains held together by simple covalent bonds, which may be broken if sufficient sheer force is applied.

The order in which the amino acid residues occur down the length of the protein is called the "primary structure". Proteins also have other structures related to their form and configuration. The angle between the atoms forming the peptide bonds allows a protein to fold into two characteristic shapes: These are the alpha helix, in which the protein spirals around a central core, and the beta pleat, in which the protein is folded in semiaccordion fashion. These are termed the "secondary structure." Tertiary structure refers to the folding of different parts of the chain, that is, how the various pleats or spirals relate to each other. Quaternary structure describes the way in which multiple chains pack together to form a molecule. All the information necessary to form a complete protein is contained within the amino acid order, that is, within the primary structure, and the conditions under which the protein is formed allow it to assume a configuration according to the secondary, tertiary, and quaternary structures: The sequence specifies the conformation.

Another important issue related to the conformation and the solubility of the protein is the formation of disulfide links. Cysteine residues contain a free sulfhydryl group in the side chain. Two cysteines in the same molecule, within close proximity, may eliminate the hydrogen atoms attached to the sulfhydryl groups and form a covalent bond between the two sulfur atoms. This lends stability and structure to the protein, while making it less soluble, for the most part, in aqueous solution.

The conformation of a protein gives it specific binding properties. It is the conformation of a protein that imparts enzymatic or drug therapeutic activity to a protein. It also provides specific binding properties between the protein and other molecules. These specific binding properties may be utilized in purification.

Nonsecreted proteins from recombinant organisms often have significant amounts of nucleic acid that is associated with the protein and must be separated from the product. The negative charge of the nucleic acid phosphate groups binds to anion exchangers.

3.2 BASES FOR SEPARATION

A tremendous variety of proteins occurs in nature. The number of distinct molecules that can be generated from twenty amino acids combined in groups of a thousand or more is immense. This variety makes the job of protein purification frustrating because there are no simple solutions: Each protein is different and will require a different process. But this variety also offers benefits, for many different methods of purification can be employed.

One of the major differences between proteins is the surface charge. This is related to the number of acidic and basic side chains. While it is not yet possible to predict from the primary structure of a protein the performance of separation processes based on surface charge, it is possible to gain some idea about how a protein may be expected to perform by virtue of its surface charge. That is, proteins that contain a large number of basic side chains would be expected, first of all, to be soluble in water and, second, to be separated from molecules that do not have a large number of basic side chains, purely through the use of this surface charge. A large number of positive groups (lysine, arginine, histidine) or negative groups (aspartate, glutamate) may suggest cation or anion exchange. However, only the groups on the surface will interact with the resin matrix.

Another basis for separation is the hydrophobicity, or relative affinity for water, of the different side groups as they make up the protein molecule. Proteins that contain a large number of amino acid residues with aliphatic and aromatic side chains are less soluble in water, and this provides an option for purification.

Another basis for separation is the size and shape of the molecules. Molecules come not only in different molecular weights but in different configurations. That is, some are globular or spherical in shape and others are long and narrow, and these differences in size and shape also form a basis for separation.

3.3 SOLUBILITY

The solubility of a protein has a major impact on its purification: Most separation methods require a soluble starting material. Many proteins produced by biotechnology are insoluble as a result either of molecular engineering (such as fusion proteins) or of high levels of overexpression within the cell (as inclusion bodies) or simply of the natural structure of the protein (with many hydrophobic or crosslinking groups). The challenge this poses is not only the solubilization of the protein for purification: In many cases the protein must be refolded and renatured to restore biological activity. Refolding may contribute significantly to the total downstream cost.

The insolubility of proteins is the result of several factors, including the presence of aliphatic and aromatic side chains in the amino acid residues and the presence of disulfide bridges between different parts of the molecule containing cysteine residues. Many insoluble proteins have economic value, but—as noted above—all purification methods require solubilization of the proteins or at least solubilization of the impurities to separate them from the product of interest.

Melander and Horvath (1977) have described the solubility of proteins in terms of a combination of salting–in (electrostatic) and salting–out (hydrophobic) effects. This is summarized in Figure 3.2, which shows the log of the solubility of the protein as a function of salt concentration. The solubility is normalized by dividing the weight of the protein soluble in 1,000 g of solution by the weight soluble in 1,000 g of pure water. The line Λ shows the salting–in effect of electrostatic interactions; the line $-\Omega\sigma_i$ shows the salting–out effect of hydrophobic interactions. K_s is a salting–out constant given by the equation

$$K_s = \Omega\sigma_i - \beta \tag{3.1}$$

The terms $\Omega\sigma_i$ represent an intrinsic salting–out coefficient. The magnitude of Ω increases with the nonpolar contact area of the protein molecules; σ_i is the molal–surface–tension increment of the salt. The magnitude of the intercept β is practically zero at the isoelectric point of the protein and increases with the net charge.

There are many mechanisms for solubilizing proteins. One is to alter the pH. Sometimes proteins that are insoluble at a low pH may be soluble at a high pH and vice versa. Many recombinant proteins are soluble in dilute alkali solutions (pH 10); an example from *Escherichia coli* is given by Lowe et al. (1987).

Another method is to add a solubilizing agent. There are three types of solubilizing agents: detergents, chaotropic agents, and reducing agents.

Detergents, like soaps, are characterized by a bifunctional molecule. That is, they have a hydrophobic end and a hydrophilic end. The hydrophobic end is often an aliphatic hydrocarbon. In the case of sodium dodecyl sulfate (SDS) the hydrophobic end is an aliphatic group (a twelve–member carbon chain), and the hydrophilic end is a strong ion (sulfonate). These agents solubilize the protein through interaction of the hydrophobic portion of the detergent with the hydrophobic portion of the molecule, thus shielding the protein from the water, while the hydrophilic end of the detergent is exposed to the water, thus allowing the protein to separate itself from the other hydrophobic proteins and move into solution. Detergents form micelles, which are small spheres consisting of a membrane of the detergent, with the hydrophobic end of the molecule inside the sphere and the hydrophilic end outside.

Many types of detergent are available, classified by their ionic nature: anionic, cationic, nonionic, and zwitterionic (bifunctional). Table 3.2 lists examples of detergents and compares their critical micelle concentration (CMC), the molecular weight of the monomer, and the apparent molecular weight of the micelle. Proper selection of the detergent is important because, as Regnier (1987) points out, not all detergents are compatible with all separation methods and in the case of chromatographic separations complete intermolecular dissociation is requisite for good resolution. In particular, ionic detergents may bind to an ion exchange column. Welling et al. (1987) suggest that mild conditions, such as physiological pH and nonionic detergents, are required when denaturation must be avoided.

Table 3.2. Comparison of detergents.*

		CMC (mM)	Monomeric MW	Micellar MW
Anionic				
	SDS (in water)	8.1	288	17,000
	SDS (0.05 M NaCl)	2.3	288	24,000
	SDS (0.5 M NaCl)	0.5	288	38,100
	Na cholate	8.3	431	2,100
	Na deoxycholate	2.7	392	9,100
Zwitterionic				
	CHAPS	4–6	615	6,150
Nonionic				
	Tween 80	0.01	1,300	76,000
	Triton X-100	0.3	650	90,000
	Lubrol PX	0.1	582	64,000
	Octyl glucoside	25.0	292	8,000

* Data from Hjelmeland et al. (1983), Hjelmeland and Chrambach (1984), Furth et al. (1984), and Welling et al. (1987).

The micelle size and critical micelle concentration of SDS are strongly dependent on salt concentration. The effective radius, known as the "Stokes radius," is proportional to the temperature and inversely proportional to the diffusion coefficient and viscosity. The bile salts (cholate and deoxycholate) have limited solubility below pH 7. SDS binds to proteins, changing their

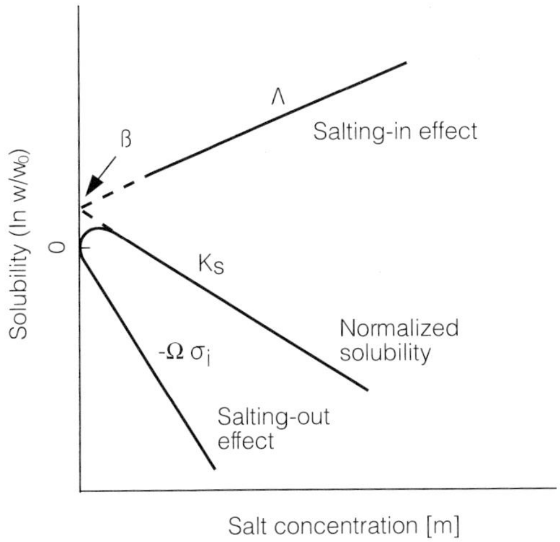

Figure 3.2. Diagram of solubility of proteins as a function of salt concentration. [From Melander and Horvath (1977).]

net charge to negative. SDS has low solubility at high salt concentration (>0.2 M) and at temperatures below 15°C. Detergents are commonly used at concentrations of 0.1 to 1 percent. The nonionic detergents tend to have larger micelles. The amount of SDS that binds to a protein is nearly constant and, according to Tanford (1980), equals 0.4 g dodecyl sulfate per gram of protein at low concentrations and 1.4 g dodecyl sulfate per gram of protein at high concentrations. The transition takes place at 6 to 7 $\times$ 10^{-4} M dodecyl sulfate at 25°C and pH 5.6 to 7.2, with little dependence on ionic strength.

Another type of solubilizing agent is the reducing agent. These agents work to reduce the disulfide bonds in the molecule, and change them into free sulfhydryl groups. Examples are 2–mercaptoethanol, dithiothreitol (DTT), dithioerythritol, glutathione, and 3–mercaptopropionate. DTT, developed by Cleland (1964), has a shorter half–life than mercaptoethanol but has a low volatility and does not have the noxious odor of mercaptoethanol. Stevens et al. (1983) describe the pH dependence of various thiol compounds in aqueous solution. At pH 6.5 and 20°C the half–life of mercaptoethanol is greater than 100 hours and that of DTT is 40 hours. At pH 8.5 and 20°C the half–lives of mercaptoethanol and DTT drop to 4 and 1.4 hours, respectively. Metal ions such as copper decrease the stability of reducing agents; EDTA increases their stability. These reducing agents also have reduced stability at elevated temperatures.

Proteins may be solubilized by the use of chaotropic agents, which tend to disrupt the order of the water molecules surrounding the protein, lessening the aversion of the protein for water. Hofmeister discovered in 1888 that salts cause the precipitation of proteins at different levels of efficiency. The Hofmeister series corresponds not only to the order in which proteins are precipitated (see Chapter 8) but also roughly to the order in which salts stabilize proteins as well as the order in which they affect hydrophobic binding (see Chapter 12). Called the "lyotropic series" (from the Greek *luein*, meaning "loosely put together"), the series consists of two parts: the kosmotropes and the chaotropes. Srinivasan and Ruckenstein (1980) have described the action of these salts in terms of their effect on the structure of water. Though the three–dimensional structure of water that results from strong hydrogen bonding is transitory, at any instant it exists to a given extent. The addition of some salts increases this structure while the addition of other salts decreases this structure. Structure–making salts are called kosmotropes (from the Greek *kosmo*, meaning "order") and structure–breaking salts are called chaotropes (from the Greek *chaos*, meaning "formless"). Kosmotropes increase the surface tension of water while chaotropes decrease the surface tension of water.

Knuth and Burgess (1987) have listed the chaotropic series for anions for protein solubilization as

$$\begin{aligned} &CBr_3COO^- > CCl_3COO^- > SCN^- > I^- > Br^- \\ &\quad > CF_3COO^- \gg CH_3COO^- > \text{citrate} > PO_4^= > SO_4^= \end{aligned} \tag{3.2}$$

and for cations as

$$\text{Guanidine} > Ca^{2+} > Li^+ > Na^+ > K^+ \tag{3.3}$$

While thiocyanate and guanidine are commonly used as chaotropes, whenever a toxic agent is added, its impact on later steps including its removal must be considered. The lyotropic series is listed in order of molal surface–tension increments in Appendix 3.1.

Organic solvents such as ethanol may help solubilize hydrophobic molecules; ionic bonds between proteins may be disrupted by high (around 0.2 to 0.5 M) salt concentrations. The combination of two agents, such as urea and CHAPS, may prove more effective than either separately.

Modifications to the protein itself are another technique for solubilization. An example is alkylation, in which the sulfhydryl group is reacted with another molecule to form a covalent bond, eliminating the potential of the sulfur group to combine with other sulfur groups and form disulfide links. A complete description of the chemistries involved is given by Means and Feeney (1971).

When purifying proteins, one must be concerned with not only removing the product but also removing the impurities, and therefore solubilization of the impurities is of interest as well as solubilization of the product molecule. The insolubility of the product may serve as a help in the purification. The product can be repeatedly washed, for example, with buffers of different salt concentration or pH, to remove DNA and soluble proteins such as proteases before solubilization of the product protein.

Many of the techniques of solubilization destroy the secondary, tertiary, and quaternary structure of the protein. This denaturation may result in loss of activity and, in some cases, must be avoided. There are occasions where the molecule may be renatured, that is, where the secondary and higher structure may be restored following denaturation. But with very large molecules this may be difficult. Lowe et al. (1987) suggest that the concentration of the protein is critical to the success of the renaturation step, as high concentrations may favor precipitation or aggregation.

The solubilizing agent may be removed before, concurrently with, or after renaturation. Ionic detergents may be removed by ion exchange. Others may be separated by gel filtration chromatography. Furth et al. (1984) suggest that a simple way to separate unbound detergent is to dilute to below the critical micelle concentration, at which point the micelles disperse to monomers. Slow dialysis, multistep dialysis (for example, from guanidine to urea to phosphate buffer), or slow dilution may result in the protein's renaturing itself. When proteins are diluted, it is better to add the buffer gradually rather than add concentrated protein to a large volume of buffer.

3.4 STABILITY

Proteins exhibit a wide range of stabilities. A process for the purification of a protein product must take into account the conditions that maintain the product in a stable form and avoid conditions that lead to protein degradation. Protein degradation can result from enzymatic cleavage by proteases, hydrolysis catalyzed by acid or base, disruption caused by shearing forces, and oxidation or reduction by active moieties. The addition of energy in the form of heat accelerates these processes. Proteins are also subject to oxidation by free radicals, peroxides, or oxygen. Antioxidants, such as ascorbic acid, may reduce autoxidation.

Another characteristic of proteins that is vital to purification is their inherent stability relative to the environment, that is, to the concentration of other molecules, to the pH, and to the temperature. Disulfide bonds spontaneously rearrange at pH > 10. Proteins with disulfide groups should be maintained at low pH to avoid rearrangement and aggregation.

Proteases, enzymes that degrade proteins, occur naturally in all organisms and are often a major cause of destruction of the product prior to purification. There are two ways to avoid protease activity: separate the proteases from the protein or inactivate the protease. Proteases are themselves proteins and separation processes may be devised to separate the product from degrading proteases. To do this quickly before major loss occurs may be challenging. An alternative is to inhibit the action of the proteases. Severe heating will destroy most proteases, and although many products are heat labile and subject to loss of activity upon heating, others may take the effect of heating without any loss in performance.

Many proteases require the presence of a divalent cation such as calcium or magnesium, in order to be active. Chelation of ions by EDTA or citric acid reduces proteolytic activity.

Trypsin and the other serine proteases are a major concern because they are found just about everywhere. PMSF (phenylmethylsulfonyl fluoride) has been found to inactivate trypsinlike proteases irreversibly. As reported by James (1978), PMSF is unstable in aqueous solution, with a half–life at pH 8 of only 35 minutes.

The destruction of proteins by strong acids and bases is a useful method of cleaning and sanitizing process equipment. Equipment that can withstand high concentrations of hydroxide may be sanitized and depyrogenated by contact with 0.5 N NaOH for one hour.

3.5 BUFFERS

Many properties of proteins, including charge and solubility, are dependent on the pH of the solution. To maintain a stable pH, various buffering agents are employed. The term buffer refers to a molecule that, when used in solution, resists changes in the hydrogen ion concentration of a solution and to the solution itself. Buffers are effective over only a narrow range, typically 1 pH unit on either side of the pK_a. Appendix 3.2 lists dissociation constants for some common biological buffers.

The concentration of a buffering agent and other salts in a buffer solution is determined by the use of the solution. For many applications, including chromatography and solubilization, low concentrations of buffer, in the range 20 to 50 mM, are sufficient. For other applications NaCl may be needed, and a common buffer that mimics physiological conditions is phosphate–buffered saline, which may have the composition 150 mM NaCl and 10 or 20 mM PO_4^-. EDTA is commonly added as a chelating agent at 1 to 5 mM. Formulas for dozens of different buffer solutions are given by William and Chase (1968).

REFERENCES

Cleland, W. W., "Dithiothreitol, a new protective reagent for SH groups," *Biochemistry* 3:480 (1964).

Furth, A. J., H. Bolton, J. Potter, and J. D. Priddle, "Separating detergent from proteins," *Methods in Enzymology* 104:318–328 (1984).

Gueffroy, D. E., "Buffers: a guide for the preparation and use of buffers in biological systems," Calbiochem–Behring, La Jolla, CA (1975).

Hjelmeland, L. M., and A. Chrambach, "Solubilization of functional membrane proteins," *Methods in Enzymology* 104:305–318 (1984).

Hjelmeland, L. M., D. W. Nebert, and J. C. Osborne, "Sulfobetaine derivatives of bile acids: nondenaturing surfactants for membrane biochemistry," *Analytical Biochemistry* 130:72–82 (1983).

James, G., "Inactivation of protease inhibitor phenylmethylsulfonyl fluoride in buffers," *Analytical Biochemistry* 86:574–579 (1978).

Kauzman, W., "Thermodynamics of unfolding," *Nature* 325:763–764 (1987).

Knuth, M. W., and R. R. Burgess, "Purification of proteins in the denatured state," in Protein Purification: Micro to Macro, R. Burgess (ed.), Alan R. Liss, New York (1987).

Lowe, P. A., S. K. Rhind, R. Sugrue, and F. A. O. Marston, "Solubilisation, refolding and purification of eukaryotic proteins expressed in *E. coli*," in Protein Purification: Micro to Macro, R. Burgess (ed.), Alan R. Liss, New York (1987).

Means, G. E., and R. E. Feeney, Chemical Modification of Proteins, Holden–Day, San Francisco (1971).

Melander, W., and C. Horvath, "Salt effects on hydrophobic interactions in precipitation and chromatography of proteins: an interpretation of the lyotropic series," *Archives of Biochemistry and Biophysics* 183:200–215 (1977).

Regnier, F. E., "Chromatography of complex protein mixtures," *Journal of Chromatography* 418:115–143 (1987).

Scopes, R. K., "Classical and modern techniques in protein purification," in Protein Purification: Micro to Macro, R. Burgess (ed.), Alan R. Liss, New York (1987).

Srinivasan, R., and E. Ruckenstein, "Role of physical forces in hydrophobic interaction chromatography," *Separation and Purification Methods* 9(2):267–370 (1980).

Stevens, R., L. Stevens, and N. C. Price, "The stabilities of various thiol compounds used in protein purifications," *Biochemical Education* 11(2):70 (1983).

Tanford, C., The Hydrophobic Effect, John Wiley & Sons, New York (1980).

Welling, G. W., R. van der Zee, and S. Welling–Wester, "Column liquid chromatography of integral membrane proteins," *Journal of Chromatography* 418:223–243 (1987).

William, C. A., and M. W. Chase, Methods in Immunology and Immunochemistry, vol. 2, Academic Press, New York (1968).

FURTHER READING

Creighton, T. E., Proteins: Structures and Molecular Properties, W. H. Freeman, New York (1984).

Heide, K., and H. G. Schwick, "Salt fractionation of immunoglobulins," in Immunochemistry, D. M. Weir (ed.), Blackwell Scientific Publications, Oxford (1978).

Schein, C. H., "Solubility as a function of protein structure and solvent components," *Bio/Technology* 8:308–317 (1990).

Schein, C. H., "Production of soluble recombinant proteins in bacteria," *Bio/Technology* 7:1141–1149 (1989).

Scopes, R. K., Protein Purification: Principles and Practice, 2nd ed., Springer Verlag, New York (1987).

Appendix 3.1 Molal–surface–tension increments of various salts. Units of surface tension are in 10^3 dyne g/cm mol.*

Salt	Surface tension
KSCN	0.45
$NaClO_3$	0.55
NH_4I	0.74
KI	0.84
NH_4NO_3	0.85
$KClO_3$	0.86
NaI	1.02
$NaNO_3$	1.06
NH_4Br	1.14
KBr	1.31
NaBr	1.32
CsI	1.39
NH_4Cl	1.39
$KClO_4$	1.4
NaCl	1.64
$CuSO_4$	1.82
K_2-tartrate	1.96
Na_2HPO_4	2.02
$MgSO_4$	2.1
$CuSO_4$	2.15
$(NH_4)_2SO_4$	2.16
$ZnSO_4$	2.27
K_2SO_4	2.58
Na_3PO_4	2.66
Na_2SO_4	2.73
$BaCl_2$	2.93
K_3-citrate	3.12
$MgCl_2$	3.16
$CaCl_2$	3.66
$K_3Fe(CN)_6$	4.34

* From Melander and Horvath (1977).

Appendix 3.2 Values of dissociation constants for common biological buffers.*

NAME	pK_a
Phosphoric acid	2.12 (pK_{a1})
Citric acid	3.06 (pK_{a1})
Formic acid	3.75
Succinic acid	4.19 (pK_{a1})
Citric acid	4.74 (pK_{a2})
Acetic acid	4.75
Citric acid	5.40 (pK_{a3})
Succinic acid	5.57 (pK_{a2})
MES	6.15
ADA	6.60
BIS-TRIS PROPANE	6.80
PIPES	6.80
ACES	6.90
Imidazole	7.00
Diethylmalonic acid	7.20
MOPS	7.20
Phosphoric acid	7.21 (pK_{a2})
TES	7.50
HEPES	7.55
HEPPS	8.00
TRICINE	8.15
Glycine amide, hydrochloride	8.20
TRIS	8.30
BICINE	8.35
Glycylglycine	8.40
Boric acid	9.24
CHES	9.50
CAPS	10.40
Phosphoric acid	12.32 (pK_{a3})

* From Gueffroy (1975).

CHAPTER 4

ANALYTICAL METHODS

There are three main applications of analytical methods in large–scale protein purification: initial characterization of the product, measurement of yield for process development and optimization, and in–process evaluation. In addition, finished products are subject to a battery of quality–control tests.

Characterization is done early in the life of the product to define the product characteristics and set specifications. This testing is often done as part of the research effort prior to process scale up. Characterization tests may also be necessary after scale up to verify that the quality of the product continues to meet the specifications.

Because the cost of production is closely tied to the process yield, we must have some indication of the yield to measure effects of process changes on product cost. The success of process development and scale up is, to a large extent, dependent on the accuracy and precision of the assays used. Careful measurement of the performance of alternative unit operations allows selection of those conditions that minimize loss and maximize purity.

As the process is validated by repetitive runs, a measure of comparison between batches is needed to monitor quality and reproducibility. After the process is implemented in the factory, testing will continue at various stages to monitor product quality and provide a quantitative measure of yield. For pharmaceuticals measurement of yield for major steps is required by regulatory agencies.

Each of these different applications requires a different type of assay. This chapter looks briefly at the assays required for each application and then concentrates on assays frequently used as part of process development and scale up.

4.1 CHARACTERIZATION

Tests for product characterization tend to be of a special nature: they are lengthy, involved, expensive, and required only a few times in the life of a product. Garnick et al. (1988) have described how regulatory agencies have set requirements for characterization of products from recombinant sources to ensure a consistent product, free of genetically induced variation, and how these regulations require complete characterization of the DNA segment itself, including the nucleotide sequence, restriction–enzyme mapping, and the history of production.

Product–characterization tests include composition, as determined by amino acid sequencing, peptide mapping, N– and C–terminal sequencing, and carbohydrate analysis; purity, to determine the nature and level of major and minor contaminating species; potency and biological activity, to show response to antibodies or antigens in vitro as well as in vivo animal testing.

Analytical techniques for protein characterization are also important for determination of physical and chemical properties of the product and the impurities as an aid in designing the purification procedure and selecting purification steps. Useful information includes description of major contaminants and physical properties of the product, such as thermal stability, isoelectric point, molecular weight, density, hydrophobicity, number of charged groups, and specific binding properties.

4.2 MATERIAL BALANCE

Yield calculations are an integral part of process development and scale up. (Process development is the design or selection of separation methods; scale up is the change in the volume of the batch size). Practically all processes for the production of products in biotechnology are batch processes. That is, unlike the oil industry, where one has a continuous flow of feed into an operating vessel, which runs twenty–four hours a day, seven days a week, month in, month out, processes in biotechnology tend to be much shorter and are carried out in a batchwise fashion, beginning with a discrete amount of starting material and ending with a given yield for a given batch size.

The typical scale up will start with a laboratory process, that is, a process that is conducted on the laboratory bench to produce quantities of the order of milligrams. The process must be scaled up to produce quantities of the order of grams or kilograms. Again, these are batch processes. Because the large–scale process handles much larger volumes than does the small–scale laboratory process, some steps useful at the laboratory scale may no longer be economical at the full production scale. Steps such as ultracentrifugation and affinity chromatography may become prohibitively expensive when one is dealing with, say, kilograms of material. Because of the expense involved in these large batches, the yield becomes an important matter, whereas it may not be critical at the laboratory scale. Higher yields mean lower cost per unit produced, and therefore yield is an important issue.

However, to determine the yield of the whole process, one must determine how much material goes into the process and how much is recovered by the process. And this applies to each of the individual steps comprising the process as well as the overall process. The easiest way to record the yield is with what is termed a mass or material balance. The material balance is a mathematical summation, an accounting, if you will, of the material that comes into a step and goes out of a step. It is shown diagrammatically in Figure 4.1. There is no generation or annihilation of material. (Presumably, in the purification of a protein, no protein is generated. Degradation of the protein may occur, and if that is a serious matter, then it must be taken into account in the material balance, but usually one tries to minimize degradation, and so it is not included in this quantitation.) Everything that goes into the process can be accounted for, either in one of the streams coming out of the process or as retained by some adsorbing mechanism within a phase in the process. Material balances can be written on both the total amount of material and on the specific protein, and both material balances are useful in scale up.

Consider, for example, an ion exchange column. One can determine the amount of total protein going onto the column. Then one looks also at the material coming out at the different stages of the process. There is the flow–through stream; one can measure the volume and protein concentration there, to determine the total amount of protein that comes out with that stream. There are elution streams; as the product is eluted from the column, the amount that comes off at different stages of elution can be measured.

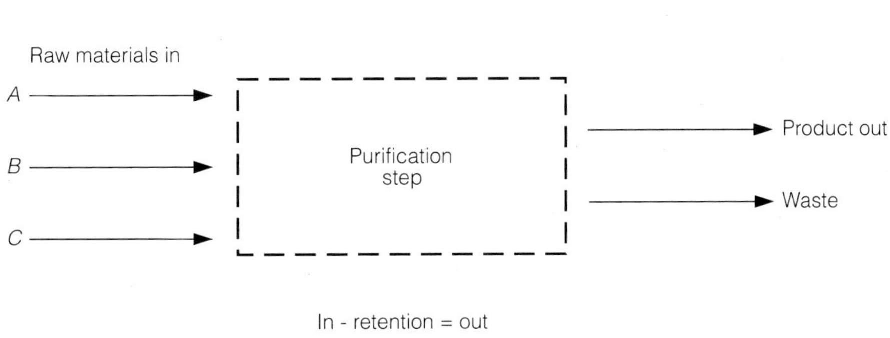

Figure 4.1. Schematic diagram of material balance over single-step or complete process.

And as the column is regenerated, the amount of material that is removed can be measured. The amount of protein in all these exiting streams, together with any that is irreversibly bound to the column, must equal the amount of the material that went into the column. All material must be accounted for.

A material balance can also be performed on the specific product protein. The amount of product going into the process in the raw–material streams, the amount coming out in each of the product streams, and the amount in each of the waste or regeneration streams can be measured. The yield of the step is the amount of protein in the product stream divided by the amount of protein in the raw–material stream. To determine this yield, we must have an assay method for determining how much is in each of these streams. This is not so difficult when we are measuring total protein but becomes more of a challenge when we look at the specific product within a mixture of various proteins and the impurities in which it is found, often at a low concentration level.

Thus tests are needed for the total protein and for the specific protein. In selecting an appropriate assay, one must consider quantitative versus qualitative assays and evaluate the different types of assays available, as discussed below.

4.3 REPRODUCIBILITY

After the optimal precess has been developed, then the reproducibility of the process must be verified, and assays developed to allow the manufacturing group to monitor the purity during the course of production. In–process assays are necessary to show reproducibility between purification lots, to monitor the cost and quality of the product, and to show variations in the process. In–process monitoring is essential to maintaining consistent quality in production.

It is also necessary that our assays be reproducible, both in terms of standardized format and precision. Accuracy (the absolute value of the result) is not quite so critical in yield determinations as the precision (relative values of multiple measurements), except at the final step. (Although knowing the actual amount is helpful in explaining the operation to upper management, strictly speaking it is not critical at each of the individual steps.)

In–process tests for monitoring usually include assays that describe the purity of the product at the various stages, assays that are used in pooling fractions from chromatography columns, and assays for measuring protein concentrations.

4.4 COMPARISONS

Though a strictly quantitative comparison of the different test methods is not possible due to the different bases of function, Table 4.1 may prove helpful in putting the various assays in perspective with regards to precision and sensitivity.

Assays are listed from left to right in order of increasing specificity, or the ability to discriminate between specific proteins, and from top to bottom in order from qualitative to quantitative. These are only relative positions for the different assays.

By light absorption, one cannot quantitatively distinguish between protein *A* and protein *B*. Size does a better job, and HPLC and immunoassays are quite good. When selecting an assay method, one would be wise to select a different basis for separation than that used in the large--scale process.

Just as in the purification process different steps are based on different physical properties for providing the separation, so also different assay methods are based on different physical properties for providing the information we require. For example, within the class of light absorbance, ultraviolet extinction, staining with Coomassie dye—as in the Bradford procedure—or staining as in the Lowry procedure are possibilities. The assays based on size include gel electrophoresis with polyacrylamide or other types of gels and various types of staining, such as Coomassie and silver. Densitometry adds some measure of quantitation to electrophoresis gels. Separation in gels is mainly due to differences in size and molecular weight and to some extent on the mobility of the molecule in an electric field. HPLC provides yet a different basis, whether ion exchange, reversed phase, or gel filtration chromatography. Immunoactivity provides another basis and includes the western blot, enzyme–linked immunoassays, radioimmunoassays, and nephelometry.

Table 4.1. Comparison of alternate test methods, illustrating relative positions of different methods with respect to sensitivity and quantitation.

	Low specificity			High specificity
	Light absorbance	Size (MW)	HPLC	Immunoreactivity
Qualitative		PAGE-Coomassie		Western blot
		PAGE-silver		
		2D gel electrophoresis		
			GFC	ELISA/RIA
		Gel densitometry	IEC	Nephelometry
Quantitative	UV extinction		RPC	

4.4.1 Tests for Total Protein Determination

Several tests are available for protein determination, but they are not specific assays: They measure all the protein present in the sample. These methods are listed in Table 4.2. The differences between the methods include the color forming agent, the pH, and in the case of the dye binding assays the part of the protein to which they bind. Because of this variation between assays, two different assays may not give the same results relative to a standard.

The confusion that might result from the use of various methods may be reduced by selecting one method, standardizing it, and using it for all yield calculations rather than trying to use multiple methods and correlate them. Of course this is not always possible because of interfering agents, and these must be taken into account when developing the method during the course of the scale up.

4.4.2 Tests for Specific Protein Determination

In addition to the total protein concentration for the overall material balance, it is also important to have a specific protein determination. These assays tend to be less precise than are the total protein methods. It is also more difficult to determine the specific protein accurately because of interference; the specific protein through much of the process is a minor component of the mixture, and other proteins may compete, increasing the background noise and decreasing the precision of the assay. Several assays, as listed in Table 4.3., are available to quantify the amount of specific protein in a mixture.

Enzyme immunoassays (EIA), also known as enzyme–linked immunosorbent assays (ELISA), are a group of methods that rely on the highly specific interaction between the product protein and either polyclonal or monoclonal antibodies. The usual arrangement has a molecule of the antigen or antibody attached to the solid phase, to which the sample binds. In the case of a competitive assay the labeled molecule competes for binding sites with the sample. In the case of the sandwich assay the sample is allowed to bind first, and then the label is brought in afterward to bind to the sample and produce a signal that can be read either by radio counting (RIA) or by colorimetric reaction.

The ELISA is a common assay that has the advantage of high sensitivity and specificity, and it can detect very small amounts of protein in a solution. It has the disadvantage that it is susceptible to background interference from

Table 4.2. Methods for total protein determination.*

Method	Basis	Binding site	Interfering agents/Comment
Lowry	Folin	Tyr, Trp, Cys, His	Amino derivatives, some buffers, reducing and chelating agents
Coomassie	Coomassie G250	Arg, His, Lys, Tyr Trp, Phe	Guanidine
OPA	O-phthalaldehyde	Lys, N-terminal	TRIS, amino acid buffers
Biuret	Copper sulfate	Coordination complex between copper and nitrogen	Requires large quantities
UV absorbance	280 nm 280/260 nm 224–236 nm 280/205 nm	Try, Trp peptide bond	Many substances absorb in this range

* See Peterson (1983) and Rickwood (1984).

Table 4.3. Methods for specific protein determination.

Method	Basis
ELISA	Immunoreaction, polyclonal or monoclonal
Nephelometry	Immunoreaction, polyclonal
HPLC	Specific sorption, chromatography
Gel electrophoresis	Electric mobility, size

competing proteins, and so, even though small amounts may be detectable, if the mixture is extremely dirty in terms of the number of other impurities, those amounts may not be reproducibly measurable. HPLC is consistent, has good precision, and can detect minor amounts of product in a complex mixture.

The rate nephelometer is a device developed for clinical chemistry assays that measure different proteins in the bloodstream. Though not often used in biotech, it is an alternative that may be found useful. A sample of antigen is added to a solution of polyclonal antibodies directed against the antigen. The immune complex that forms is sufficiently large to deflect light in a spectrophotometer. The scattered light is measured by a photomultiplier tube and the protein concentration determined by comparison with a standard curve. Beckman, Behring, and other manufacturers make automated machines.

Rate nephelometry, like the ELISA, is based on immunoreaction. Where an ELISA may be based on reaction to either polyclonal or monoclonal antibodies or to a specific antigen in the case of tests for antibodies, nephelometry requires a polyclonal antibody. It is a simple matter to determine the concentration of various samples relative to a standard curve. Nephelometric assays are quicker and simpler to run than are ELISAs, but the lower limit of detection for nephelometry is about 10 μg/ml, compared to 0.1 μg/ml for an ELISA. A thorough review of nephelometric assays for proteins is given by Whicher et al. (1983).

Of all the specific determination methods, HPLC is probably the best developed in terms of the number of applications and the number of variations in test configurations (for example, solvents, columns, and gradients). Snyder et al. (1988) and Regnier and Gooding (1980) describe detailed conditions for protein applications. Analytical HPLC commonly uses reversed phase columns, which offer high resolution of low–molecular–weight (less than 10 kD) proteins. Larger molecules and insoluble molecules may be amenable to gel filtration or ion exchange methods (see later chapters for discussion of large–scale chromatographic techniques). HPLC is a powerful technique for the quantitative analysis of proteins.

Gel electrophoresis is most useful in determining the relative purity of a sample. Gels may be made from several types of polymers, such as acrylamide and agarose. Polyacrylamide gels containing SDS in the running buffer, perhaps the most common type for electrophoresis of proteins, are also called Laemmli gels after the work of Laemmli (1970). The use of a densitometer to scan the gel and quantify the amount of the different proteins requires standardized procedures in terms of length of staining time, length of destaining time, amount of protein loaded, and so on, because of the variabilities in the stain absorption. For example, Coomassie blue, a commonly used stain, preferentially binds to arginine. To get a reproducible value from gel densitometry, one must be consistent in one's method; electrophoresis is useful as an in–process assay when it is carried out in a consistent way. It is less useful in the development of a process for determination of the yield at each step. An example of an SDS–reduced polyacrylamide gel of fractions from a gel filtration column is shown in Figure 4.2. Wilson (1983) has compared various stains for proteins in electrophoresis gels. The ratio of the size of the band and the amount of protein it represents is constant over only a small range of protein loadings. Gels can be used to estimate protein concentration only for protein loads that fall within this linear range.

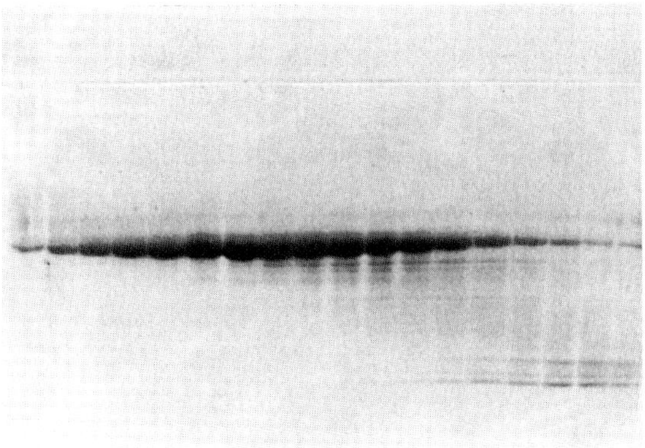

Figure 4.2. SDS-reduced polyacrylamide gel electrophoresis of fractions collected from gel filtration chromatography. (Courtesy of T. Zuffi, Chiron Corporation.)

Several methods of ultraviolet light absorption are frequently used, as listed in Table 4.4. Chromatography columns are monitored at 214, 254, 280, or 405 nm. Because nucleic acids absorb light in the same region and contribute to some error in these readings, various methods have been proposed as compensation, such as taking a ratio of readings at 260 and 280 or taking the difference between readings at 224 and 236.

Analytical isoelectric focusing is usually performed in horizontal–slab polyacrylamide gels. Precast gels are available commercially; alternatively a gradient mixer can be employed to deliver an ampholyte gradient as the gel is poored. Marker proteins of known pI can be used to indicate the position of different pH values. In this method, samples are run in discrete lanes, just as in Laemmli electrophoresis.

Table 4.4. Methods of UV absorption.

Wavelength (nm)	Absorbing species	Application
280	Tyr, Trp, nucleic acid	Protein
260/280	Nucleic acid, protein	Protein in presence of nucleic acid
236–224	Tyr, Trp, Phe, His, Met, Cys, peptide bond	Protein in presence of nucleic acid

Isoelectric focusing is also used for electrophoretic titration. Gels are made in a special casting frame that includes a trough across the center for sample application. The pH gradient is formed by application of current to the sides of the gel perpendicular to the sample trough. The gel is then rotated 90 degrees with respect to the electrodes and the sample is applied to the center trough. Under the influence of an electric current the sample migrates towards the anode or cathode a distance dependent on its net charge. As the pH varies continuously along the length of the trough, after migration the protein will form a continuous curve from one edge of the gel to the other. Electrophoretic titration curves are helpful in selecting ion exchange conditions. The pH value that gives maximum separation between multiple protein bands on a gel will often provide high resolution in ion exchange chromatography.

REFERENCES

Garnick, R. L., N. J. Solli, and P. A. Papa, "The role of quality control in biotechnology: an analytical perspective," *Analytical Chemistry* 60:2546–2557 (1988).

Laemmli, U. K., "Cleavage of structural proteins during the assembly of the head of bacteriophage T4," *Nature* 227:680–685 (1970).

Peterson, G. L., "Determination of total protein," *Methods in Enzymology* 91:95–119 (1983).

Regnier, F. E., and K. M. Gooding, "High–performance liquid chromatography of proteins," *Analytical Biochemistry* 103:1–25 (1980).

Rickwood, D., Centrifugation: A Practical Approach, IRL Press, Oxford (1984).

Snyder, L. R., J. L. Glajch, J. J. Kirkland, Practical HPLC Method Development, John Wiley & Sons, New York (1988).

Wicher, J. T., C. P. Price, and K. Spencer, "Immunonephelometric and immunoturbidimetric assays for proteins," *CRC Critical Reviews in Clinical Laboratory Sciences* 18(3):213–260 (1983).

Wilson, C. M., "Staining of proteins on gels: comparisons of dyes and procedures," *Methods in Enzymology* 91:236–247 (1983).

FURTHER READING

Dunbar, B. S., Two–dimensional Electrophoresis and Immunological Techniques, Plenum Press, New York (1987).

Dux, J. P., Handbook of Quality Assurance for the Analytical Chemistry Laboratory, Van Nostrand Reinhold, New York (1986).

Ishikawa, E., T. Kawai, K. Miyai, Enzyme Immunoassay, Igaku–Shoin, Tokyo (1981).

Kemeny, D. M., and S. J. Challacombe, ELISA and Other Solid Phase Immunoassays, John Wiley & Sons, New York (1988).

Laas, T., "Isoelectric focusing in gels," in Protein Purification: Principles, High Resolution Methods, and Applications, J.–C. Janson and L. Ryden (eds.), VCH Publishers, New York (1989).

Scopes, R. K., Protein Purification: Principles and Practice, 2nd ed., Springer Verlag, New York (1987).

Tijssen, P., Practice and Theory of Enzyme Immunoassays, Elsevier, Amsterdam (1985).

Williamson, A. R., "Isoelectric focusing of immunoglobulins," in Immunochemistry, D. M. Weir (ed.), Blackwell Scientific Publications, Oxford (1978).

CHAPTER 5

IMPACT OF UPSTREAM OPERATIONS

The production of biologicals is often split—and quite naturally so—between the so–called upstream and downstream operations. The upstream side includes fermentation and cell culture; it is the production of the desired molecule in some raw form. The downstream side includes recovery and purification of this product. Choices made in the design of the upstream side affect the design of the downstream side.

The integration of upstream and downstream operations during the design stage is often taken for granted in theory but neglected in fact. Integration may result in a product with lower cost and higher quality, thus providing a competitive edge.

It is worthwhile also to consider the impact of improvements in the upstream and downstream sides of the process. The most skillful improvement of a downstream recovery process will produce at best a tenfold increase in yield (from, say, 8 to 80 percent; it can never be over 100 percent). However, skillful improvement of the upstream fermentation process through selection of high–producing organisms, high–growth media, and optimal operating conditions may produce a thousandfold increase in production.

5.1 UPSTREAM INFLUENCES

Interaction between upstream and downstream sides of an operation occurs superficially at the point of transfer: The final product of the upstream operation is the raw material for the downstream operation. But in a broader sense interaction extends much further. For not only is the product material delivered to the downstream group, so are all the impurities. Whereas the upstream group is assigned to produce the product in a crude form, the downstream group must remove the impurities, including those added at upstream stages.

There are four variable factors in the upstream operation that affect the downstream operation: namely, the volume of material delivered, the amount of impurities relative to the product, the nature of the impurities, and the phases present (solid, liquid, or gas).

The initial step in downstream processing is usually a volume–reduction step to concentrate the product and reduce the cost of vessels needed for processing. Reduction of the volume or mass of material produced upstream, excepting product, will reduce downstream costs. This may be as simple as direct connection of the fermenter to a continuous–flow centrifuge instead of holding the fluid in a storage vessel or as complex as substitution of a cell–culture perfusion system for roller bottles. In both cases the volume leaving the upstream stage has been reduced, and smaller–volume processing systems can be used downstream.

Opportunities often exist for reducing the amount of impurities relative to product. In the culture of recombinant organisms longer growth times may lead to greater production of impurities relative to product. This may increase the difficulty of purification more than it improves the upstream yield, and thus it may have a negative impact on the product cost. Conversely, the use of a membrane bioreactor may allow the use of less expensive media (for example, newborn–calf serum instead of fetal–bovine serum) without increasing contaminants (such as immunoglobulins) and represents an upstream change with a positive effect on the downstream operation.

The nature of the impurity may be controlled at the upstream stage. One choice often considered in efforts to lower fermentation costs is to use more crude forms of media components, such as unrefined sugars and starches with municipal water in place of defined media with ultrapure water. Variations in the composition of such ingredients, between lots and throughout the year, will lead to variations in the biochemical pathways followed by the organisms, which in turn leads to the production of different impurities. Some purification procedures can handle variations in the number and nature of impurities without sacrifice of the quality of the product. Others cannot endure the slightest deviation. Thus, if the fermentation includes ingredients with high variability, then the purification must consider this if production crises are to be avoided.

The number of phases present may complicate or simplify the downstream recovery process. In one case a product secreted by the cell may be more easily recovered by ultrafiltration from simple media than from the complex mixture of broken cells; in another case whole cells containing nonsecreted product may be more easily processed than tank–car loads of dilute media.

Another method by which the downstream purification may be greatly helped by upstream activities is the purposeful alteration of the protein to include a region that aids in the recovery. This may be through fusion of an "affinity tail," as described by Uhlen et al. (1988), with the protein recovered by affinity chromatography, or it may be the incorporation of multiple charges, such as the polyarginine tail described by Smith et al. (1984), with purification by ion exchange chromatography or polyelectrolyte precipitation.

5.2 DIFFERENCES BETWEEN UPSTREAM AND DOWNSTREAM UNIT OPERATIONS

Many differences exist between upstream and downstream operations. For background consider first the various sources of biologicals or the means by which we produce these molecules, as listed in Table 5.1. In the first column are fermentation processes for the growth of various organisms: yeast, bacteria, fungi, and others. Recombinant proteins, for example, are products made this way. In the second column is cell culture: in mammalian cells, insect cells, or other cells. Monoclonal antibodies are produced this way. In the third column are animal and vegetable sources: blood products, organs, urine, plants, and other sources. Hormones for diagnostic assays may be obtained from these sources.

Upstream processes can be further classified by the scale at which they are used, as shown in Table 5.2. Fermentation covers a wide range, with some products produced at a scale of less than 10 l per batch and others produced at over 10,000 l per batch. Cell–culture systems—roller bottles and perfusion processes—tend to be conducted at a smaller scale, usually less than 100 l per batch, although occasionally more. Extraction processes from blood and urine may be hundreds of liters per batch, while organ extracts are typically of the order of tens of liters per batch.

Very–large–scale fermentations (>10,000 l) are accompanied by recovery processes that can handle these volumes. Flow–through centrifuges and filtration are continuous processes that can handle virtually any volume. These unit operations effectively reduce the volume of the product to a more manageable scale for subsequent purification operations, such as

Table 5.1. Sources of proteins.

Fermentation	Cell culture	Animal/vegetable
Yeast	Mammalian	Blood
Bacteria	Insect	Organs
Fungi	Other	Urine
Other		Plants
		Other

Table 5.2. Unit operations classified by scale.

	less than 100 l	100–1000 l	More than 1000 l
Production			
Fermentation	×	×	×
Roller bottles	×	×	
Perfusion	×	×	
Extraction	×	×	
Recovery			
Centrifugation	×	×	×
Filtration	×	×	×
Chromatography	×	×	
Electrokinetics	×		

chromatography, which are less costly at smaller scale. Thus one of the key differences between upstream and downstream processes is the scale of operation. Some of the other differences between upstream and downstream operations are outlined in Table 5.3.

Table 5.3. Differences between upstream and downstream processes.

Upstream	Downstream
Live organisms	No organisms
Large volumes	Progressively smaller volumes
Many components	Progressively fewer components
Heterogeneous mixture	Homogenous solution
Capital intensive	Labor intensive
Automated	Manual
Few operations	Many operations
Lower cost materials	Higher cost materials

These differences fall into three groups. First, downstream operations rarely involve live organisms. In the cases of fermentation and cell culture products the growth of microorganisms is perhaps the most crucial stage of the upstream process. Opportunity exists here to increase expression, use lower cost materials, and in many ways reduce the cost or increase the yield of the operation. The concern of the downstream side, however, with regard to cells is how to break them open to recover an intracellular product or how to remove them from the media in the case of a secreted product.

Second, downstream processes work at progressively smaller scales in terms of volume, number of components, and number of phases. The upstream side involves large volumes containing many diverse components in a heterogeneous mixture of cells and media. Challenges include mass–transfer problems and the efficient mixing of three phase systems containing gas, liquid media, and solid cells. At progressive stages of the downstream process the volume decreases, the number of components diminishes, and the solution becomes homogeneous. The mixture is filtered, clarified, and concentrated. The thermodynamic efficiency of each step is more important in downstream operations.

Third, upstream and downstream processes differ in terms of cost factors. Upstream operations tend to be capital intensive, with large–volume automated vessels, few operations, and relatively lower–cost materials. Downstream operations, on the other hand, tend to be labor intensive, with little automation, many operations, and higher–cost materials.

Similarities also exist, and these similarities provide opportunities for synergy in the design.

5.3 SYNERGISTIC OPPORTUNITIES

The first similarity between the upstream and the downstream is obviously the product. We said earlier that the upstream side of the operation was concerned with generating the product and the downstream side was concerned with recovering the product. Through both operations the product usually remains unchanged. What varies is the *environment* in which the product is found. For both upstream and downstream operations the product has value and is the focus of attention. Both sides have the same goal of producing a high–quality product at the lowest cost, and the process is designed to this end. System design and operation design (see Chapter 1) require joint effort between the upstream and downstream design groups. Specification of the complete system requires expertise in both areas, as does deciding which types of operation to include.

Beyond consideration of the product there are other areas of overlap between the upstream and downstream design groups. An example is process control. As mentioned earlier, one of the differences between upstream and downstream operations is the cost factor. The combination of high capital cost and few operations on the upstream side has led to highly automated systems, with controls and monitors at every step. Downstream processes tend to consist of multiple manual operations; and although it may not be cost effective to automate the downstream operation fully, connection of monitoring devices to the data–collection system provides an inexpensive means of tracking process conditions. Figure 5.1 illustrates the vertically integrated process, tied together by process monitoring. The product flows

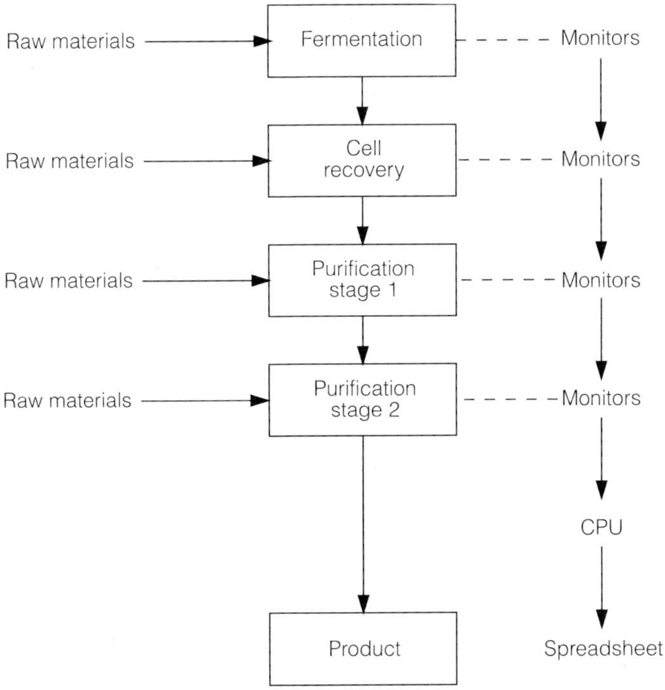

Figure 5.1. Flowchart of a vertically integrated process.

downward through the center stages, raw materials are added at each stage, and process data is collected at each stage and assembled in a spreadsheet. The information provided by these monitors is especially useful in troubleshooting production problems. The expertise in control theory of the upstream designers can provide benefits to the downstream group through integration of the process into a cohesive whole.

Another aspect of cooperation between upstream and downstream technologies is the incorporation of recovery steps into the upstream cycle. Figure 5.2 illustrates a potential design in which the high selectivity of affinity chromatography is utilized to recover the product in a semi–continuous manner. Media from the culture vessel (hollow fiber, stirred tank, or other) is pumped through an adsorbing column for a time, after which the product is eluted and the column returned to the collection mode. The advantage of such an arrangement is rapid recovery and separation of the product from culture conditions. This may be especially important if the product is liable to rapid degradation or if the product inhibits expression.

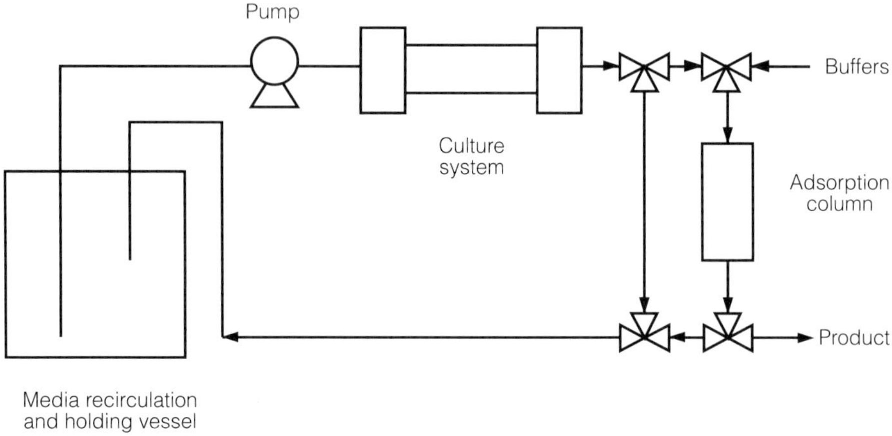

Figure 5.2. Schematic diagram of integrated recovery and production system.

Many differences exist between the upstream and downstream operations, but opportunities for synergistic cooperation also exist. Both sides are concerned with production of a common product at the lowest cost. The influence of the upstream operation on the cost of the downstream recovery shows how joint consideration of the system design is required to produce the optimal process. And beyond the overall system design the specification of the process steps benefits from the analytical skills of the detached group. The whole operation can be tied together by data collection and process control.

The decision to integrate the upstream and downstream design can have a significant impact on the final cost of the product. Upstream and downstream groups often work separately, with little contact. This may seem to be cost saving because it avoids time–consuming meetings, but it is false economy. By saving a little, we may pay a lot, with a suboptimal process and a high potential for later production problems. Vertical integration of upstream and downstream processing results in optimization of the whole operation rather than of independent parts.

REFERENCES

Smith, J. C., R. B. Derbyshire, E. Cook, L. Dunthorne, J. Viney, S. J. Brewer, H. M. Sassenfeld, and L. D. Bell, "Chemical synthesis and cloning of a poly(arginine)–coding gene fragment designed to aid polypeptide purification," *Gene* 32:321–327 (1984).

Uhlen, M., T. Moks, and L. Abrahmsen, "Protein engineering to optimize recombinant protein purification," *Biochemical Society Transactions* 16(2):111–112 (1988).

FURTHER READING

Bejurstrom, E., "Biotechnology: fermentation and downstream processing," *Chemical Engineering* 92(4):126 (1985).

Germino, J., and D. Bastia, "Rapid purification of a cloned gene product by genetic fusion and site–specific proteolysis," *Proceedings of the National Academy of Sciences, U. S.* 81:4692–4696 (1984).

Reisman, H.B., Economic Analysis of Fermentation Processes, CRC Press, Boca Raton, FL (1989).

CHAPTER 6

CELL DISRUPTION

Product recovery often requires cell disruption: The product may be soluble but found within the cell, or the product may be insoluble. Many overproduced proteins are found clumped together in inclusion bodies, small nodules of insoluble protein segregated within the cell. These nonsecreted soluble proteins and insoluble proteins, whether they are inclusion bodies found in yeast or bacteria or cell–membrane proteins, must be separated from the cellular components before they can be purified.

Cell–disruption methods can be categorized into four classes: nonmechanical methods, ultrasonics, high pressure homogenization, and mechanical grinding. Each of these has characteristics that make it favored under different conditions. The latter two are most amenable to large–scale application in modern biotechnology.

In the simplest case the desired product is in solution within the organism, and simple fracture of the cell wall is sufficient to release the product. Situations arise, however, where the cell must be more thoroughly disrupted to allow complete product recovery.

A wide variety of cell–wall structures exist in nature, ranging from very fragile to very tough. The more fragile the structure, the less shearing force that must be applied to break open the cell. Mammalian cells tend to be shear sensitive and require only a minimal energy input for disruption. This poses a significant problem if one is trying to pump or otherwise handle such cells but greatly simplifies the task of product recovery. Animal tissue cells may be easily disrupted by changing the osmolarity, by the action of detergents, or by application of low levels of ultrasonic energy.

Yeast and other forms of fungi have rigid cell walls containing polysaccharides. For disruption the strength of these cell walls requires the drastic action of freeze pressing or the high shear forces of bead milling or high–pressure homogenization. Of the bacteria gram–positive forms tend to have more rigid walls than do gram–negative forms, and the amount of shear necessary to disrupt bacterial cell walls varies widely between forms. The high–pressure homogenizer and the ball mill, originally designed for, and still widely used in, milk and paint production, have proved efficient in the large–scale disruption of many types of yeast, bacteria, and other microorganisms.

The criteria by which one selects the alternative disruption methods include the amount of shear allowable for the product, the specificity of the method, the need to control temperature or energy input, and the capital cost.

6.1 QUANTITATION

To the extent that recovery is dependent on the number of cells disrupted or the size of the resultant particles, some measure of the extent of breakage is required. Microscopic measurements are most common although other techniques are also available. The particles one deals with in a disruption step are typically of the order of a few microns in size. Table 6.1 gives dimensions of a few examples.

The simplest microscopic determination of the number or fraction of broken cells is simple visual estimation. Living cells and broken or dead cells can be more easily distinguished from each other if phase–contrast optics are used, where the live cells appear as bright spots of light and the dead or broken cells appear as dark shadows. A ruled reticle may aid estimation.

Table 6.1. Diameters and densities* of common organisms and cell particles.

Particle	Diameter (μm)	Density (g/cm^3)
DNA		1.7
Soluble proteins	0.001–0.01	1.3
Virus	0.01–0.1	1.07
Nuclei	3–12	1.1–1.3
Mitochondrion	0.5–4	1.1–1.2
Bacteria	0.5–5	
E. coli	1–2	1.05–1.1
Yeast	2–15	1.05–1.2
S. cerevisiae	4–5	1.1
Red blood cells	5–8	1.1
Plant cells	20–30	
Human liver cell	30–40	
Human egg cell	60	

Resolution of light microscope is approximately 0.5 μm.

* Densities based on data of Lloyd and Poole (1979) and Rickwood (1984).

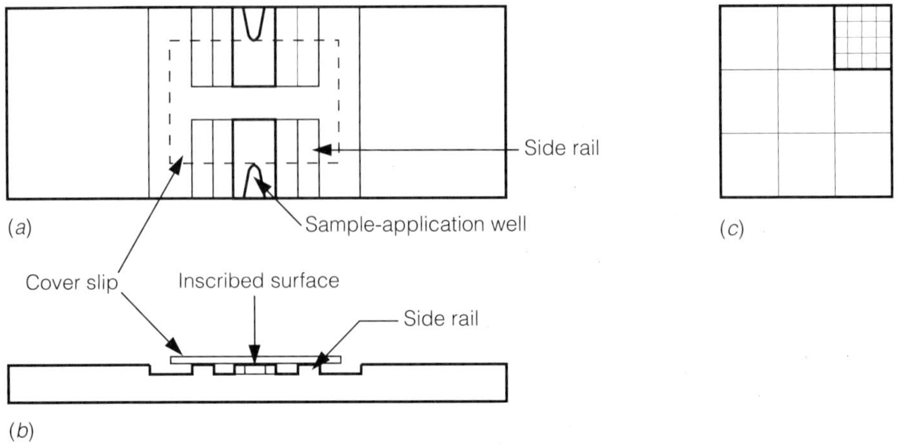

Figure 6.1. Diagram of hemacytometer, or counting slide: (*a*) top view, (*b*) front view, (*c*) detail of incised grid, showing the sixteen small squares within one corner square. Volume of outlined square is 5×10^{-5} ml.

A more quantitative value can be obtained by the use of a hemacytometer, or counting slide. Originally devised for counting blood cells, the hemacytometer provides a constant–volume chamber underlain by a precisely inscribed grid, as shown in Figure 6.1. The cover slip rests on side rails, slightly above the slide surface. Capillary attraction holds the fluid sample in contact with the top and bottom of the liquid chamber, providing a constant volume. The volume of liquid above each of the nine major squares is 5×10^{-5} ml. The grid lines provide a way to count the exact number of cells in each area. In practice the count of several squares is usually averaged. The number of cells per unit volume is the concentration; the number of cells in one of the nine major squares multiplied by 20,000 gives the number of cells per milliliter. From measurements taken before and after (as well as during) breakage the fraction of cells disrupted can be determined.

The particle size may also be determined by sieving, counting in a cell sorter, or by some sort of fineness gauge, such as the Hegman type, although the latter is difficult to use with particles less than 5 μm in diameter.

6.2 NONMECHANICAL DISRUPTION

The high shear forces of the mechanical disrupter may lead to denaturation of proteins or disruption of desired organelles. Among the alternatives for nonmechanical disruption of cells are osmotic shock, freezing, detergents, and enzymes. In each case the cell membrane may be either totally disrupted or made partially permeable to allow the product to escape.

Osmotic shock is the sudden introduction of cells into a solution of low osmolarity, such as the transfer from 25–percent sucrose solution into pure water. If the change is rapid, ions inside the cell may not be able to migrate through the cell wall before the osmotic pressure differential causes rupture of the cell. In a similar way slow freezing rates, which promote large crystal growth, may disrupt cells. Cells may be made permeable by the use of detergents (such as sodium dodecyl sulfate, SDS) to solubilize the lipid bilayer and in some cases disrupt the cells.

In the case of enzymatic lysis the cell wall is the substrate for enzyme activity. The commercially available enzymes for cell lysis are limited to lysozyme made from egg whites, applicable to only a few bacteria (such as *E. coli*). An advantage of enzymatic lysis, compared with other disruption methods, is better control of pH and temperature, lower capital investment, and potential specificity, while a disadvantage is the expense of enzymes. Enzymes may be used in connection with mechanical methods to increase efficiency and reduce power requirements. Andrews and Asenjo (1987) suggest that enzymes may be used to release only specific proteins, thus simplifying the downstream operation and reducing the total recovery cost, compared with nonspecific homogenization methods.

Nonmechanical methods of cell disruption do not disrupt the DNA of the cells. DNA from ruptured cells polymerizes and greatly increases the viscosity of the solution. Digestive enzymes may be used to degrade the nucleic acid and reduce the viscosity. Mechanical methods may disrupt the DNA at the time the cells are broken.

6.3 ULTRASONICS

Ultrasonic disruption occurs when sound waves having a frequency of the order of 20,000 cps (20 kHz) are converted to very rapid vibration in a liquid, causing a phenomenon known as cavitation. The very rapid vibration of the ultrasonic device causes local low–pressure areas within the liquid. The pressure in these areas is low enough to convert the liquid to gas in the form of very small bubbles. As the local pressure changes again and begins to rise, these bubbles collapse. The collapse of these bubbles causes a shock wave to travel through the liquid, resulting in a shear force that will disrupt cells.

The ultrasonic energy is transmitted to the liquid through a "sonicator" (or "sonifier") tip, or "horn." As White and Marcus (1988) explain, the amplitude of the vibration at the end of the horn is inversely proportional to the diameter of the tip; large–diameter tips deliver less sonic energy than smaller tips do. The rapid vibration of the tip generates a large amount of heat that must be dissipated by turning the sonicator off every few minutes during use. Coakley et al. (1977) recommend power densities of 0.2 to 3.0 W/ml; higher power densities result in severe local cavitation at the tip, which reduces the effectiveness of the operation.

The use of ultrasonic power for large–scale disruption is limited by the output that is possible from a sonicator tip as well as by the large amount of heat that is generated. Doulah (1981) has suggested the use of a bank of sonifiers in series, but it is doubtful that such a system would compete economically with other large–scale methods.

6.4 HIGH–PRESSURE HOMOGENIZATION

High–pressure homogenizers generate a rapidly moving liquid stream that impinges against either an immovable surface or a second high–pressure stream. Impingement against a surface is used in a device originally described by French in a paper by Milner et al. (1950) and subsequently called a "French Press," which consists of a piston for applying high pressure to the cell suspension and forcing it through a narrow orifice.

The critical components of this device are the valve and valve seat, shown in Figure 6.2. The valve consists of a hardened rod that is brought into close contact with the valve seat. The cell suspension is forced through the central hole of the valve seat, where it impinges against the valve and is forced around the valve past the impact ring into a low–pressure chamber.

The operating pressure is adjusted by altering the distance between the valve and valve seat. Various designs for the valve and valve seat are available, as shown in Figure 6.3. The internal beveled valve seat (cell disrupter) and the flat valve seat (standard valve) are commonly used for microorganism disruption.

Variation in the concentration of cells in the feed suspension of 450 to 750 g/l has been shown by Hetherington et al. (1971) not to affect the disruption process, whereas at higher cell concentrations the breakage efficiency decreases. Doulah et al. (1975) suggest that at a given pressure the cells at the upper end of the size distribution will break before smaller cells break. The higher the pressure, the greater the fraction of cells that will be disrupted. Pressures of 35,000 to 140,000 kp (5000 to 20,000 psi), with corresponding velocities of 180 to 280 m/s (600 to 900 ft/s), are common for cell breakage. At higher pressures the amount of protein recovered in a

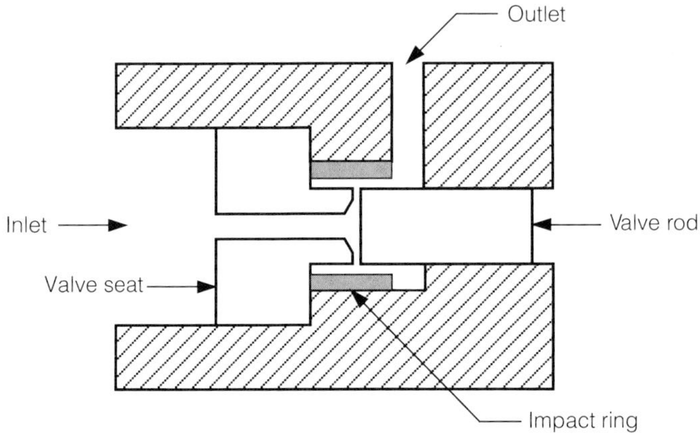

Figure 6.2. Diagram of valve and valve-seat placement for high-pressure homogenizer. (Reproduced with permission of APV Gaulin, Inc.)

single pass through the homogenizer increases dramatically. Recycling the product through the homogenizer is an alternative to higher pressures, but with several disadvantages: Throughput of the equipment will be decreased, and wear on the valve seat will be increased by the amount of recycle; released products are subject to degradation from longer processing times and repeated subjection to the high temperature generated in the disruption valve; multiple passes reduce the size of debris particles, making their later separation more difficult.

The amount of disruption that can be obtained at a given pressure is very dependent on the type of organism being disrupted. Organisms with stronger

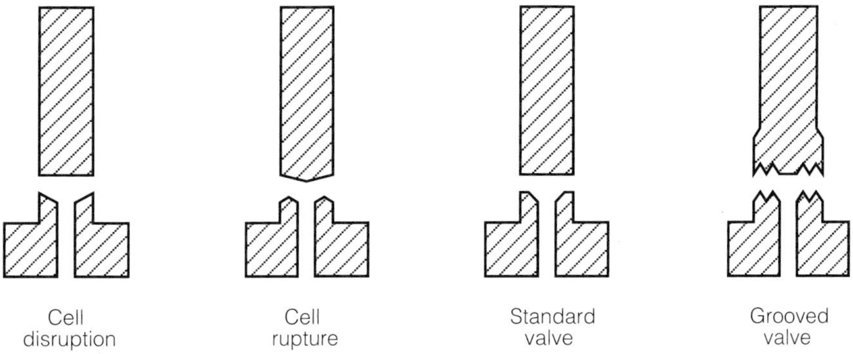

Figure 6.3. Diagram of various valve and valve-seat configurations used in microorganism disruption. (Reproduced with permission of APV Gaulin, Inc.)

walls will require higher pressure to achieve a given level of breakage. Breakage efficiencies of 90 percent or higher usually require multiple passes through the valve. High–pressure homogenizers add considerable heat to the process stream, and some type of cooling is frequently required, especially with heat–labile products.

Another style of high–pressure homogenizer is the impinging–jet fluidizer, shown in Figure 6.4. In this device the process stream is divided in microchannels, accelerated to velocities up to 450 m/s (1,500 ft/s) and directed against an opposing stream rather than against a metal valve. Operating pressures for cell disruption are similar to those of the valved device, namely, 35,000 to 70,000 kp (5000 to 10,000 psi).

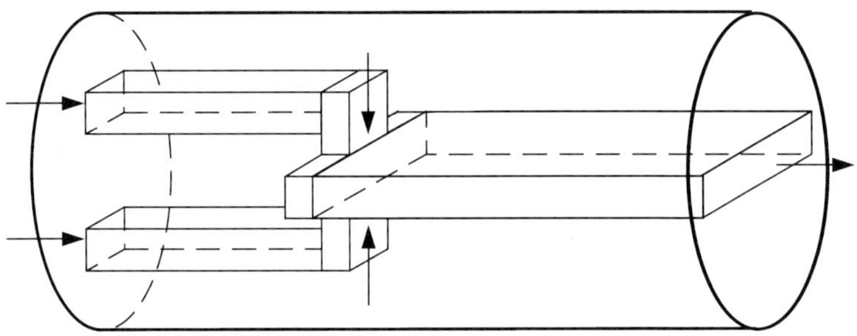

Figure 6.4. Diagram of impaction device for impinging jet fluidizer. The process stream is divided into two jets that impinge against each other at high velocities. (Courtesy of George Nikolopoulos, Microfluidics Corporation.)

The mechanism of cell disruption in these devices is still disputed, but the three most likely candidates are cavitation, turbulence, and shearing. Evidence exists in support of each of these mechanisms, and while one may predominate under certain circumstances, it is probable that all three contribute to the total disruption process.

Fluid impingers of up to 75 l/m (20 gpm) and valve homogenizers of up to 95 l/m (25 gpm) are available commercially. Both designs suffer occasional blockage of the orifice. Laboratory models that handle samples as small as 120 ml are available.

When a high–pressure disrupter is scaled up, it is important to maintain the exit velocity (pressure), the orifice and valve–seat design, the throughput or flow rate, the recycle rate, the temperature of operation, and the fluid viscosity.

6.5 MECHANICAL GRINDING

Mechanical grinders consist of a cylindrical chamber, containing cells in liquid suspension, and grinding media, such as glass beads, with a turning shaft. In one design a shaft with multiple disks spaced at regular intervals along its length turns at high speed, causing the beads to impact against the cells, which leads to disruption. Alternatively, the grinder may consist of a rolling or vibrating chamber with the beads thrown violently against the sides, impacting against the cells and causing disruption. Disruption in mechanical grinders results from impact and shear force applied against the cell wall by the grinding media. Figure 6.5 illustrates the major parts of a horizontal bead mill. Disruption occurs in a controllable manner in a completely enclosed system. Both horizontal and vertical chambers are available; the primary difference in operation is the grinding–media load or ratio of grinding–media volume to chamber volume, 50 to 60 percent in vertical machines and 80 to 85 percent in horizontal machines. Horizontal mills have more grinding points per unit volume—and thus higher grinding efficiency—as well as higher friction and greater temperature rise.

The major parameters of concern for the scale up of mechanical grinders are the grinding media and the shaft speed, or peripheral velocity. Glass beads are typically used for grinding cells although beads of other materials, such as steel or ceramic, may also be employed. Uniformly sized grinding media, 0.1 to 2 mm in diameter, are preferred for consistent results. Bacteria require smaller beads, typically on the order of 0.1 mm diameter, while yeast require slightly larger beads, around 0.5 mm in diameter. A smaller bead size tends to be more effective because of the increased number of contact

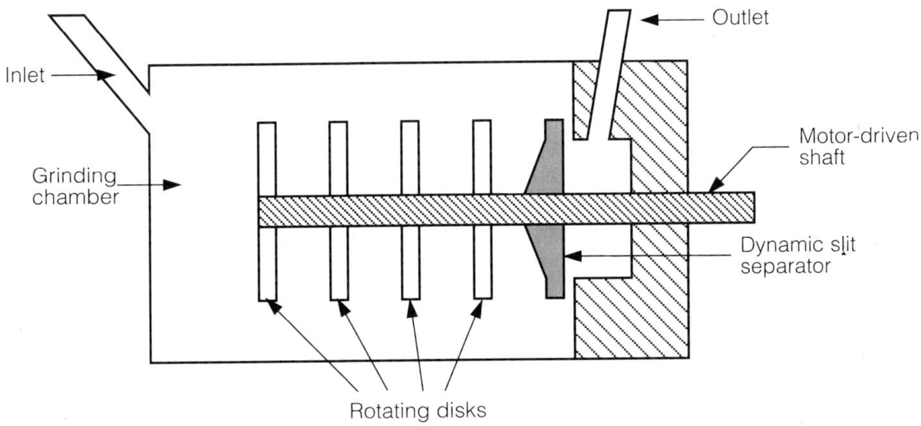

Figure 6.5. Diagram of horizontal bead mill, showing major parts.

points between the grinding media and the work to be milled. The optimal media size is determined by the material to be ground; the media size remains constant regardless of grinding–chamber volume. The grinding chamber is filled with grinding media before introduction of the cell slurry. Overfilling with media generates excessive wear due to limited clearance between the agitator disks and the chamber wall.

The rate at which the agitator shaft rotates influences the rate at which the grinding beads impact the cells, and therefore a higher shaft speed typically leads to more rapid disruption. In scale up the peripheral velocity is used rather than the shaft speed because the optimal shaft speed depends on the diameter of the grinding chamber, while the optimal peripheral velocity is independent of vessel size. The peripheral velocity that can be achieved depends on the viscosity of the material, the type of grinding media, and the type of agitator disk. At high peripheral velocities the wear on the grinding media and the increase in temperature can be very high. Typical working peripheral velocities are on the order of 10 to 15 m/s. Schutte et al. (1983) reported that, for grinding baker's yeast, as the tip speed increased, the power consumption increased at a rate faster than the rate of protein release.

Several other factors bear on the efficiency of the disruption process as measured by the rate of disruption and the fineness of the particles generated: the throughput or flow rate, the viscosity, the temperature, the configuration of the grinder (that is, spacing between disks along the rotating shaft), and the ratio of grinding media to total volume.

As in the high–pressure homogenizer, increasing the flow rate in the mechanical grinder decreases the particle size or the fraction of cells disrupted in a single pass through the device. To increase the fraction of cells disrupted, one may decrease the flow rate or recycle the stream through the grinder. In addition the first one and one–half chamber volumes to pass through the mill are usually returned to the feed. It is possible in the mechanical grinder to increase the flow rate to a point at which the grinding media are displaced by convective flow toward the outlet end, where it gets held up by the separator and results in a higher concentration or density of grinding media at one end of the vessel than at the other. This high concentration of grinding media may lead to a greatly increased viscosity and a subsequent temperature rise. The convective movement of the grinding media can be counteracted by raising the shaft or peripheral velocity, using larger grinding media, using media having a higher specific gravity, or reducing the viscosity of the solution being ground. Limon–Lason et al. (1979) found that an increase in the longitudinal mixing or back flow led to a decrease in efficiency, especially at low flow rates.

The viscosity of the slurry being ground affects the grinding efficiency in several ways. As mentioned, higher–viscosity fluids contribute to convective movement of the glass beads toward the outlet end of the grinding chamber, leading to decreased efficiency. On the other hand, increasing the viscosity to as high as 100 cp may improve the efficiency of breakage because of generation of higher shear forces. For microorganism disruption feed–stream compositions of 12 to 17 percent dry weight (1/1 dilution of packed–cell paste) are common. The viscosity also influences the wear of the glass beads and machine parts: The higher the viscosity, the less the wear, and therefore dense solutions contribute to longer life of equipment and grinding media. While most cell suspensions show newtonian or slightly thixotropic behavior, that is, decreased viscosity with increased shear, some solutions may be rheopectic and become more viscous as the peripheral velocity or shear increases.

The high–energy input of a mechanical disrupter typically leads to a high temperature increase, and therefore some type of cooling is necessary when heat–labile materials, such as proteins, are involved. The shaft bearings and seals also require cooling. Refrigeration systems using ethylene glycol are usually sufficient to keep the operating temperature around 4°C. Low operating temperatures increase the viscosity of the grinding solution and increase the friability of the cells, leading to more efficient disruption.

Each manufacturer recommends his own style of disk, with variations occurring in shape and eccentricity around the shaft. The mechanical energy of the rotating shaft is imparted to the grinding media through the impeller disks. Specially designed disks counter the compression of grinding media toward the discharge end in viscous solutions. Limon–Lason et al. (1979) suggest that the rate of disruption is a function of not only spacer shape but also the number of impeller disks per unit volume.

The grinding rate is proportional to the number of contact points among the grinding media per unit of volume, and therefore more concentrated grinding media in the disrupter will lead to a higher grinding disruption efficiency. In some cases the minimum size of the grinding media is limited by the mechanism used to separate the disrupted cells leaving the chamber from the grinding media. Two types of device are commonly employed: One is a straight sieve, which may have some type of vibration or crossflow to flush the grinding media away and keep the sieve clear; the second type, referred to as a dynamic slit, is a small aperture formed between a disk on the rotating shaft and a disk mounted to the grinding vessel. The dimensions of the slit are easily adjusted to conditions of grinding–media size and fluid viscosity such that the grinding media cannot pass through (typically one–third of the smallest grinding element used). The rotation of the disk attached to the agitator shaft works to keep the opening free of blockage.

Large–scale units typically use a slotted screen, with openings small enough to prevent passage of grinding media. Bead–mill grinders handle flow rates up to about 10 gpm. Laboratory models are available that handle as little as 150 ml per batch.

The optimal type of cell disruption depends on the application and the scale. In a study by Seva et al. (1986), comparing the impinging jet fluidizer, the valve homogenizer, and the horizontal bead mill, 95–percent breakage of *Saccharomyces cerevisiae* required thirty passes in the impinging–jet fluidizer compared with a 3.3–minute residence in the horizontal bead mill. A 90–percent breakage of *E. coli* required one pass through the impinging–jet fluidizer compared with two passes through the valve homogenizer. For breakage of *Bacillus subtilis* three passes through the valve homogenizer gave 90–percent breakage compared with only 55–percent breakage after three passes through the impinging–jet fluidizer. Other studies have shown similar differences between equipment.

REFERENCES

Andrews, B. A., and J. A. Asenjo, "Enzymatic lysis and disruption of microbial cells," *TIBTECH* 5(10):273–277 (1987).

Anonymous, "Cell disruption with APV Gaulin high–pressure homogenizers," *Technical Bulletin 74*, APV Gaulin, Everett, MA (1985).

Coakley, W. T., A. J. Bater, and D. Lloyd, "Disruption of microorganisms," *Advances in Microbial Physiology* 16:279–341 (1977).

Doulah, M. S., "Cell disintegration by ultrasonic cavitation," *Process Biochemistry* 16:16 (1981).

Doulah, M. S., T. H. Hammond, and J. S. G. Brookman, "A hydrodynamic mechanism for the disintegration of *Saccharomyces cerevisiae* in an industrial homogenizer," *Biotechnology and Bioengineering* 17:845–858 (1975).

Hetherington, P. J., M. Follows, P. Dunnill, and M. D. Lilly, "Release of protein from baker's east (*Saccharomyces cerevisiae*) by disruption in an industrial homogenizer," *Transactions of the Institute of Chemical Engineers* 49:142–148 (1971).

Limon–Lason, J., M. Hoare, C. B. Orsborn, D. J. Doyle, and P. Dunnill, "Reactor properties of a high speed bead mill for microbial cell rupture," *Biotechnology and Bioengineering* 21:745–774 (1979).

Lloyd, D., and R. K. Poole, "Centrifugation—principles and instrumentation," *Techniques in Life Science* B202:1–27 (1979).

Milner, H. W., N. S. Lawrence, and C. S. French, *Science* 111:633 (1950).

Rickwood, D., Centrifugation: A Practical Approach, IRL Press, Oxford (1984).

Schutte, H., K. H. Kroner, H. Hustedt, and M. R. Kula, "Experiences with a 20 liter industrial bead mill for the disruption of microorganisms," *Enzyme and Microbial Technology* 5:143–148 (1983).

Seva, R., J. Fieschko, R. Sachdev, and M. Mann, "Cell breakage using a microfluidizer," paper presented at Annual Meeting, Society of Industrial Microbiology, San Francisco (1986).

White, M. D., and D. Marcus, "Disintegration of Microorganisms," in Downstream Processes: Equipment and Techniques, A. Mizrahi ed., Alan R. Liss, New York (1988).

FURTHER READING

Chisti, Y., and M. Moo–Young, "Disruption of microbial cells for intracellular products," *Enzyme and Microbial Technology* 8(4):194–204 (1986).

Leuthold, A., "Parameters and interactions of various factors involved in dyno–mill grinding technology," Glen Mills, Maywood, NJ.

MacNeill, C., S. R. Sneeringer, and A. Szatmary, "Optimization of cell disruption with bead mill," *Pharmaceutical Engineering* 5(6):34–38 (1985).

Schieritz, M., "Disintegration of microorganisms under laboratory and production conditions," Willy A. Bachofen AG Maschinenfabrik, Basel.

The Hegman fineness gauge is described in ASTM D 1210.

APPENDIX 6 Vendors of large–scale disruption equipment.

Vendor	Equipment
APV Gaulin 44 Garden St Everett, MA 02149 617–387–9300	High–pressure homogenizer, valve impingement
Glen Mills Inc. 203 Brookdale St. Maywood, N.J. 07607 201–845–4665	United States distributor for Willy A. Bachofen, Horizontal bead mill grinder
Willy A. Bachofen AG Utengasse 15/17 CH–4005 Basel Switzerland 061/33 55	Horizontal bead mill grinder
Microfluidics Corp 44 Mechanic Street Newton, MA 02164 617–965–7255	High–pressure homogenizer, liquid jet impingement
Netzsch Inc. 119 Pickering Way Exton, PA 19341 215–363–8010	Vertical bead mill grinder
Rannie a/s 8, Roholmsvej DK–2620 Albertslund Denmark 45 2 64 93 00	High–pressure homogenizer

CHAPTER 7

CENTRIFUGATION

Centrifugation is a unit operation that finds great utility in downstream processing, with its widest application in the separation of solids from liquids. Centrifuges come in many sizes, with various designs for the removal of particles from whole cells to entities as small as 0.5 μm. Centrifuges may also be used for the classification or separation of different sized particles within a liquid medium.

Two methods of centrifugation are available to the process designer: rate (nonequilibrium) sedimentation and isopycnic (equilibrium) sedimentation. In rate sedimentation a particle travels from its initial location in the rotor until it reaches the outer wall or until the operation is interrupted. If the particle remains in the rotor for a sufficiently long time, it will reach the outer wall of the rotor.

In isopycnic centrifugation the particle travels toward the rotor edge until it reaches a point in the medium where the densities of particle and medium are equal, and here it remains, regardless of the length of time of centrifugation after it reaches this point. Strict solid–liquid separations, the most common application of centrifuges in biotechnology, are all of the rate–sedimentation type. Isopycnic separations are used for the isolation of small particles, such as viruses, and require a fluid medium in the rotor that varies in density from the inner to the outer edge.

Centrifuges come in many designs. Ultracentrifuges have small capacity and are run at forces exceeding 100,000*g*. Flow through centrifuges come in three types—stacked disks in bowls, concentric chambers, and long cylinders—and handle throughputs up to 150,000 l/h at forces less than 20,000*g*. Pilot–scale centrifuges operate at lower throughput rates and higher forces. Laboratory centrifuges handle volumes up to 6 l per cycle, at forces up to about 20,000*g*.

7.1 DESIGN PRINCIPLES

A particle spinning in a centrifuge is subjected to a centrifugal accelerating field or centrifugal force field in a direction perpendicular to, and away from, the axis of rotation,

$$F_c = \omega^2 r \tag{7.1}$$

where ω is the angular momentum and r is the distance from the axis of rotation. A more convenient measure is the relative centrifugal force (RCF), or the centrifugal acceleration relative to the earth's gravitational acceleration:

$$\mathrm{RCF} = \frac{\omega^2 r}{g} \tag{7.2}$$

A particle subjected to a relative centrifugal force of $2g$ would settle at twice the rate of a particle subjected only to gravity. In practice, the speed of rotation is more conveniently used than the angular velocity. One revolution is equal to 2π radians; so the angular velocity is

$$\omega = \frac{2\pi N}{60} \tag{7.3}$$

where the N is the speed in revolutions per minute (rpm) and the units of ω are inverse seconds. The earth's gravitational acceleration is 9.81 m/s^2; so for r in meters

$$\mathrm{RCF} = (1.12 \times 10^{-3}) r N^2 \tag{7.4}$$

For example, a particle spinning at 10,000 rpm, 0.1 m from the axis of rotation, would have a relative centrifugal force of 11,200g, and a particle spinning at 3,000 rpm, 0.4 m from the axis of rotation, would have a relative centrifugal force of 4,032g. This relationship provides a way to relate the centrifugal force in different types and sizes of rotors. Unfortunately the literature sometimes does not recognize this distinction, and centrifugation data are given in terms of the rotor speed without information on the rotor diameter. From Equation 7.4 it is obvious that a laboratory rotor of 4 cm radius and an industrial rotor of 40 cm radius subject a particle to significantly different forces.

The distance from the axis of rotation to the particle varies with time as the particle moves toward the edge of the rotor, and the relative centrifugal force increases near the outer wall of the rotor. The relative centrifugal force is often calculated for the center of the rotor or the outer wall of the rotor. For tube–type centrifuges the change in radius is small, but for disk–type centrifuges the change in radius is significant. Appendix 7.1 shows a nomograph for computing relative forces.

The time required for a particle to sediment depends on the sum of the forces acting on it. In addition to a centrifugal force particles in a centrifuge are also subject to buoyancy and drag forces acting in the opposite direction from the centrifugal sedimenting force. The total force is the sum of the centrifugal, buoyant, and drag forces, and it is equal to the mass of the particle times its acceleration, or

$$F_c - F_b - F_d = m\frac{dv}{dt} \tag{7.5}$$

The centrifugal force is the centrifugal acceleration times the mass, or

$$F_c = \omega^2 rm \tag{7.6}$$

The buoyant force is given by Archimedes's principle as the mass of the fluid displaced by the solid particle times the acceleration, or

$$F_b = \omega^2 r\left(\frac{m}{\rho_p}\right)\rho_m \tag{7.7}$$

where ρ_p is the density of the particle and ρ_m is the density of the fluid medium. The drag force results from friction between the particle and the surrounding medium, and is given by Hsu (1981) as

$$F_d = \frac{1}{2}(C_d A \rho_p v^2) \tag{7.8}$$

where C_d is a drag coefficient (defined below), A is the cross–sectional area of the particle normal to the direction of movement, and v is the velocity of the particle through the fluid. Substitution of Equations 7.6, 7.7, and 7.8 into Equation 7.5 gives

$$\frac{dv}{dt} = \left(1 - \frac{\rho_m}{\rho_p}\right)\omega^2 r - \left(\frac{C_d A \rho_p v^2}{2\,m}\right) \tag{7.9}$$

The drag coefficients for the flow regimes of most interest are given by Kunii and Levenspiel (1969) as

$$C_{\mathrm{d}} = \frac{24}{\mathrm{Re}} \quad \mathrm{Re} < 0.4\,(\text{Stokes region}) \tag{7.10}$$

$$C_{\mathrm{d}} = \left(\frac{10}{\mathrm{Re}}\right)^{1/2} \quad 0.4 < \mathrm{Re} < 500\,(\text{Allen region}) \tag{7.11}$$

where Re is the Reynold's number, $\rho_p v d_p/\mu$, and d_p is the diameter of the particle and μ is the fluid viscosity. Substitution of Equations 7.10 and 7.11 into 7.9, along with $A = \pi d_p^2/4$ and $m = \rho_p \pi d_p^2/6$ and the assumption that the acceleration is zero, gives the terminal velocity

$$v = \frac{(\rho_p - \rho_m)\,\omega^2 r\, d_p^2}{18\,\mu} \quad \mathrm{Re} < 0.4 \tag{7.12}$$

$$v = \left[\frac{0.18\,(\rho_p - \rho_m)^2 \omega^4 r^2 d_p^3}{\rho_p \mu}\right]^{1/3} \quad 0.4 < \mathrm{Re} < 500 \tag{7.13}$$

For example, a yeast cell of density 1.1 g/cm^3 and 20 μm diameter sedimenting in media of density 1.05 g/cm^3 and viscosity 1.5 cp (1 g/m s, typical for a 2–percent, weight–per–weight suspension), in a rotor of 4 cm radius spinning at 5,000 rpm has a terminal velocity of 0.8 cm/s and a Reynold's number of 0.12 and travels from the center of the rotor to the outer edge in 5 s. Note the effect of the particle size, which is raised to the second power. Though proteins are denser than cells, the contribution of this difference is only a factor of 2 to 10, while the contribution of the size difference is a factor on the order of 10^6 (see Table 6.1). Thus particles are easily separated from proteins even though the density of the latter is greater.

Equation 7.12 represents the common form of Stokes's law. Note that the settling velocity is proportional to the square of the particle size and inversely proportional to the viscosity. In narrow channels and solutions of high solid content particle–wall and particle–particle interactions will lead to deviations from this ideal. Particle recovery in a centrifuge can be increased by reducing the settling distance or by increasing the sedimentation velocity through manipulation of the variables in Equations 7.12 and 7.13. The density of the liquid medium, as well as its viscosity, may be reduced by dilution. The particle size may be increased by flocculation or aggregation. The viscosity increase that results from the release of cellular nucleic acid may be reduced by the action of nuclease or by dilution.

In actual practice the acceleration is not zero because the relative centrifugal force increases with increasing distance from the rotor. However, a sedimentation time calculated from the terminal velocity will represent the maximum time in a fixed–volume rotor. If the fluid is moving through the rotor, the trajectory will not be a straight line from the center to the outer edge but will follow a more complex path. Other effects, such as the Coriolis force, also play a role. The complexity of the flow in a centrifuge rotor necessitates empirical studies for scale up.

The relative ease with which a particle sediments is a useful measure for comparing different particles. The sedimentation coefficient defines the behavior of a particle in terms of the ratio of its velocity to its centrifugal acceleration,

$$S = \frac{\nu}{\omega^2 r} \tag{7.14}$$

The unit of S is the Svedberg, defined as 10^{-13} s (for most biological macromolecules the magnitude of S is about 10^{-13} s). The value of the sedimentation coefficient depends on the density and viscosity of the fluid medium and therefore must be determined for the particular conditions of a separation. Diameters and densities for some common biological materials are listed in Table 6.1.

The centrifuge was developed as an improvement upon early gravity settlers, and a brief description of a settler will aid in understanding centrifuge development. A gravity settler, as shown in Figure 7.1, consists of a large tank with feed at flow rate Q and area A. Particles that settle at least as far as the depth of the moving layer will be trapped in the stagnant layer. The maximum velocity at which a given particle can travel and still be trapped is equal to the flow rate Q divided by the area A. Similarly, the maximum velocity through a centrifuge for the collection of a given particle size is the flow rate Q divided by the effective rotor area A and the relative centrifugal force, or

$$Q = \nu A(\text{RCF}) \tag{7.15}$$

The equivalent clarification area of a rotor, Σ, is the product of the effective rotor area and the relative centrifugal force

$$\Sigma = A(\text{RCF}) \tag{7.16}$$

Σ represents the area needed in a gravity settling tank to equal the efficiency of a centrifuge. Flow–through centrifuges have been developed so as to maximize Σ, the equivalent clarification area. The Σ factor has proved

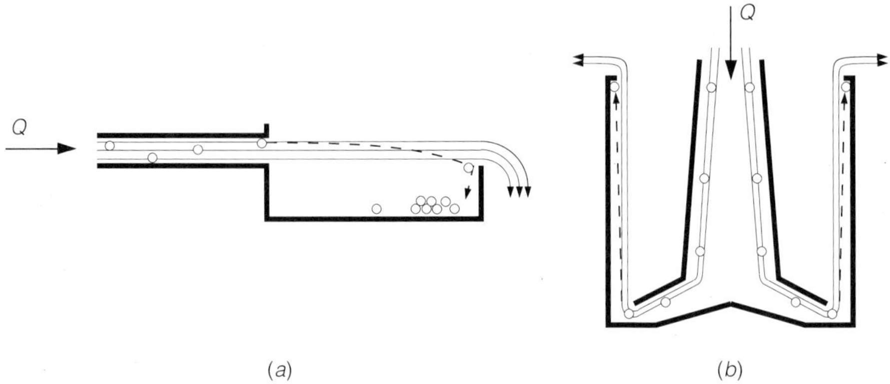

Figure 7.1. Comparison of gravity and centrifugal settlers.

especially valuable in scale up, in spite of nonperfect flow patterns in real processes: Mosqueira et al. (1981) found good correlation between pilot–scale centrifuges and laboratory results.

Three rotor designs have evolved for flow–through centrifuges: the disk type, the tube type, and the chamber type, as illustrated in Figure 7.2. In addition, fixed–volume (laboratory) centrifuges are also used in biotechnology. A comparison of rotor types is given in Table 7.1.

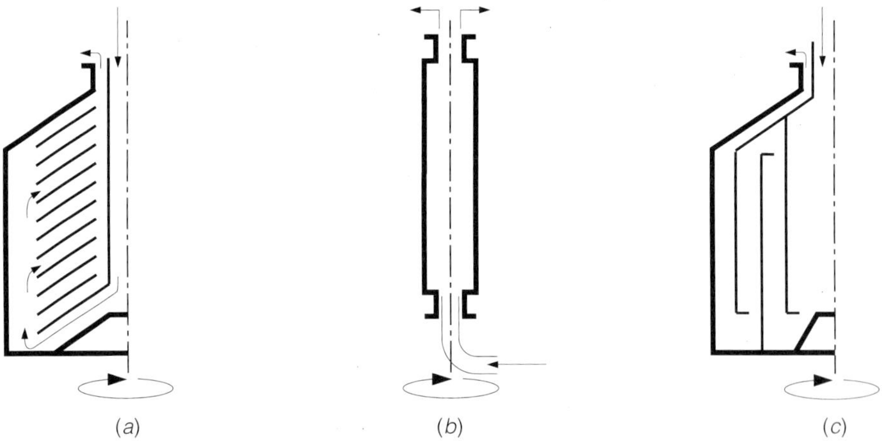

Figure 7.2. Three designs for flow-through rotors: (*a*) disk, (*b*) tube, (*c*) chamber.

Table 7.1. Comparison of centrifuge rotor types.*

Class	Speed (RPM × 0.001)	Diameter (cm)	RCF (g × 0.001)	Maximun Flowrate (l/hr)	Σ (m^2 × 0.001)	Maximum Capacity (l)
Pilot scale						
disk	8–12	20–30	6–8	500	3–22	0.5
tube	30–50	2–5	40–60	8	0.5–2	0.5
ultra	48–55	12–15	150–170	8	16–22	4
Industrial scale						
disk	3–7	16–50	7–10	150,000	20–200	60
tube	13–18	7–15	13–17	600	1.5–4	10
chamber	5–10	12–50	6–11	15,000	10–50	65
ultra	35	13	90	80	28	8
Laboratory Scale						
bottles	2–20	15–30	5–20	10	NA	6

* All centrifuges listed are flow-through models except for the laboratory centrifuge, which requires bottles.

7.2 FLOW–THROUGH CENTRIFUGES

7.2.1 Disk–type Rotor

The largest–volume centrifuges consist of a spinning bowl containing parallel stacked plates. The solids are discharged from the bowl in either a periodic or a continuous manner. The feed enters the bowl near the bottom, and the supernatant, or clarified liquid, is removed from the top. The separated solids are collected in the outer perimeter of the bowl.

An important consideration in the development of centrifugal–separation equipment has been the economic value of short sedimentation times. In the disk–type centrifuge this is accomplished through the use of a stack of conical disks that divides the entire sedimentation volume into thin flow channels.

The flow rate Q through a disk–type rotor can be described by substitution into Equation 7.15 of Stokes's relationship for the velocity (Equation 7.12) and various assumptions about the flow through the thin channel between disks; see Murkes (1966) or Hemfort (1984):

$$Q = \left[\frac{(\rho_{\mathrm{p}} - \rho_{\mathrm{m}})\, d_{\mathrm{p}}^2}{18\ \mu}\right]\left[\frac{2\pi}{3}\right]\left[\frac{\omega^2 z\,(r_1^3 - r_2^3)}{\tan\ \varphi}\right] \tag{7.17}$$

where z is the number of individual separating spaces, r_1 is the outer radius, r_2 is the inner radius, and φ is the angle of the separating space from the horizontal, as shown in Figure 7.3.

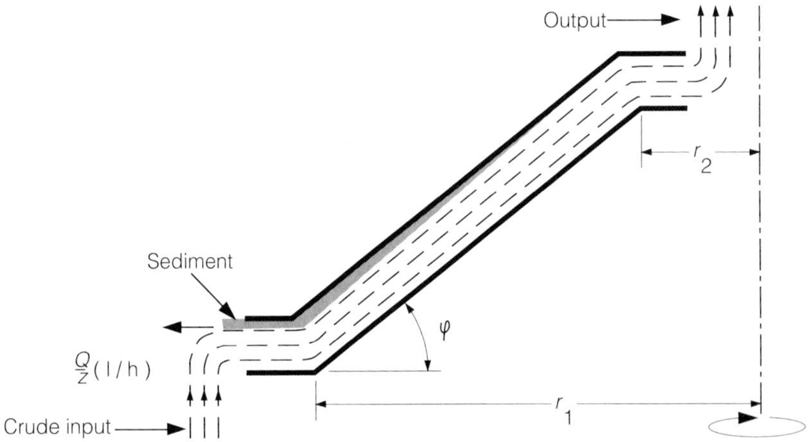

Figure 7.3. Pattern of flow between parallel plates in disk-type centrifuge.

Table 7.2. Example separation of *Escherichia coli* in Westfalia model CSA8 centrifuge.*

Culture characteristics	Centrifuge characteristics
Smallest cell size: $d_p = 0.8 \times 10^4$ cm	Bowl speed: N = 9200 rpm
Cell density: $\rho_p = 1.05$ g/cm^3	Angular velocity: $\omega = 963$ s^{-1}
Medium density: $\rho_m = 1.02$ g/cm^3	Number of disks: z = 72
Viscosity: $\mu = 1.02 \times 10^{-2}$ g/cm/sec	Disk angle: $\varphi = 38°$
Gravitational acceleration: g = 9.81 m/sec^2	Inner radius: $r_2 = 3.6$ cm
	Outer radius: $r_1 = 8.1$ cm
Q = 327 L/hr	

* From Brunner and Hemfort (1988). Brunner and Hemfort found that in actual practice flow rates were 200 to 400 l/h, indicating that Equation 7.17 is a good first approximation.

The first term of Equation 7.17 describes the physical properties of the suspension and is equal to the sedimentation coefficient, the velocity divided by the acceleration. The last term describes the relevant properties of the separator bowl. Thus the flow of a particular separation is proportional to the sedimentation coefficient and the rotor separating capacity. As described by Murkes (1966), workers at Alfa–Laval have made corrections in the exponents of r_1 and r_2 to 1.5 and 2.75, respectively, to better represent empirical measurements. The minimum particle that can be sedimented at a particular flow rate is given by Equation 7.17. For a given rotor and suspension all factors are constant, and $Q \propto dp^2$. Thus any increase in the throughput leads to a decrease in the maximum particle size that is retained. An example is shown in Table 7.2. Longer residence times lead to capture of smaller particles, with the drawback that heating of the process fluid occurs at residence times exceeding one bowl volume per minute.

The solids may be removed from the bowl in various manners. The continuous desludger has a nozzle that is always open, and the centrifugal force generated by the spinning bowl expels the sludge into a catch basin. The periodic desludger has a valve on the nozzle that opens to allow the expulsion of solids into the catch basin. The valve may be actuated by a timing device or by the monitoring of the turbidity of the clarified liquid. The bowl may be partially desludged without interruption of the feed stream. Complete desludging requires the shut off of the feed to prevent loss of liquid with the solids. Discharge of solids usually requires only a few seconds. The continuous–discharge nozzle delivers a solid slurry with a higher water content than that of the periodic desludger. Many disk–type centrifuges can be cleaned in place and steam sterilized. The most common material for parts contacting product is stainless steel.

7.2.2 Tube–type Rotor

The tube–type rotor consists of a long narrow cylinder with the feed at one end (typically the bottom) and the clarified–liquid exit at the other end. Unlike the disk–type centrifuge, which attains high–g forces at low speeds because of its large diameter, the tube–type rotor must spin at high speeds to attain the equivalent g force. The smaller mass of the cylinder relative to the disk rotor allows high–speed operation without a substantial difference in power load.

Tube–type rotors require manual removal of solids. They are widely used when a high relative centrifugal force is required, as in the sedimentation of small particles, and batch sizes are small. The solids in a tube–type rotor can be spun to a higher degree of dryness than in a disk rotor, where the solids must flow through the discharge nozzle.

7.2.3 Chamber–type Rotor

The chamber type rotor, as can be seen from Figure 7.2, is a solid–wall bowl centrifuge with high capacity but manual solids recovery. The chamber rotor is also used for very–high–speed (ultra) centrifugation, as discussed in the next section. The chamber consists of a series of concentric cylinders that provides a short settling path and long liquid residence times, each cylinder operating as an individual tube–type rotor. Recovery of the solids requires disassembly of the chamber and manual scraping of the bowls.

While the chamber–type rotor has high solids capacity, the manual solids recovery is a disadvantage in some operations. It is mainly employed in the polishing of solutions with low solids content and the in handling of sensitive proteins that cannot be subjected to the shear associated with automatic–discharge units.

7.3 ISOPYCNIC SEDIMENTATION AND ULTRACENTRIFUGATION

Sedimentation of particles by density requires a variation in the density of the liquid within the centrifuge. This density gradient may be continuous or discontinuous and may be preformed or self–forming. Equilibrium sedimentation may be carried out in either fixed–volume or flow–through centrifuges. The flow–through ultracentrifuge differs from the flow–through clarifier in that in the ultracentrifuge the feed solution is not continuously added to the rotor; rather it is more of a batch process. Once the gradient is formed, a limited amount of sample is applied, spun, and fractionated before more sample is applied.

Table 7.3. Media used in isopycnic gradients.

Media class	Example	Density (g/cm^3)	Self-forming
Sugar	Sucrose	1.0–1.6	No
Polysaccharides	Ficoll	1.0–1.6	No
Colloidal silica	Percoll	1.1–1.2	Yes
Metal salts	Cesium chloride	1.0–1.9	Yes

The density of a continuous gradient varies smoothly from one edge to the other. Discontinuous gradients have sharp changes in density. Preformed gradients are set before the sample is applied. Self–forming gradients may be generated as the sample and fluid are spun together. Preformed gradients are prepared by filling the rotor with medium of decreasing density, by use of a gradient former (a mixing chamber for combining continuously varying proportions of light and heavy fluids) or by stepwise addition of solutions of different density. In the fixed–volume rotor this is done manually to each container; in the flow–through rotor the solution is pumped in, usually while the rotor is spinning at low speed.

Self–forming gradients may be formed before or after the sample is added. Such salts as cesium chloride or such colloids as polyvinylpyrrolidone–coated silica (Percoll) will form a continuous gradient upon centrifugation. Table 7.3 lists several gradient–forming media.

Large–scale ultracentrifuges are available with both tube–type rotors and single–bowl chamber rotors. The Electro–Nucleonics Model K is an example of the former, with flow rates up to 80 l/h and forces up to 90,000*g*; the Beckman flow–through ultra is an example of the latter, with flow rates up to 3 l/h and forces up to 250,000*g*.

In both cases the gradient is established at low speeds, and the speed is increased to the selected operating level. The feed enters at the bottom, particles such as virus move into the gradient, and the less–dense supernatant is removed from the top of the rotor. Once the gradient is saturated, it must be displaced from the rotor with a dense solution. For collection of the desired product the gradient may be fractionated as it leaves the rotor. Isopycnic separations in two different rotors are equivalent when the centrifugal effect, defined as the integral of the square of the angular velocity ω over time, is kept constant.

7.4 FIXED–VOLUME CENTRIFUGES

Fixed–volume centrifuges, also known as laboratory centrifuges, use bottles or tubes to contain the sample as it spins in the rotor. The larger units hold up to 6 l and operate at forces up to 20,000*g*. The centrifuging ability of a laboratory centrifuge can be expressed in terms of the product of *g* force and capacity in liters; standard models range from 10,000 to 30,000.

REFERENCES

Brunner, K. H., and H. Hemfort, "Centrifugal separation in biotechnological processes," in Downstream Processes: Equipment and Techniques, A. Mizrahi (ed.), Alan R. Liss, New York (1988).

Hemfort, H., Separators, Westfalia Separator AG, Oelde, Federal Republic of Germany (1984).

Kunii, D., and O. Levenspiel, Fluidization Engineering, John Wiley & Sons, New York (1969).

Hsu, H. W., Separations by Centrifugal Phenomena, John Wiley & Sons, New York (1981).

Mosqueira, F. G., J. J. Higgins, P. Dunnill, and M. D. Lilly, "Characteristics of mechanically disrupted bakers; yeast in relation to its separation in industrial centrifuges," *Biotechnology and Bioengineering* 23:335–343 (1981).

Murkes, J., "Outline of the theory of separation processes in centrifugal separators," *Filtration & Separation* 3:112–114 (1966).

FURTHER READING

Bond, P., A. Scragg, and M. Fowler, "Recovery and disruption of plant cells," in Separations for Biotechnology, M. S. Verrall and M. J. Hudson (eds.), Ellis Horwood, Chichester, England (1987).

Lloyd, D., and R. K. Poole, "Centrifugation—principles and instrumentation," *Techniques in Life Science* B202:1–27 (1979).

Miller, S. A., C. M. Ambler, R. C. Bennett, D. H. Dahlstrom, J. D. Darji, R. C. Emmett, J. B. Gray, C. F. Gurnham, L. J. Jacobs, R. P. Klepper, A. W. Michalson, J. Y. Oldshue, C. E. Silverblatt, J. C. Smith, and D. B. Todd, "Liquid–solid systems," in Chemical Engineers' Handbook, 6th ed., R. H. Perry, D. W. Green, and J. A. Maloney (eds.), McGraw–Hill, New York (1973).

Rickwood, D., Centrifugation: A Practical Approach, IRL Press, Oxford (1984).

APPENDIX 7.1. Nomograph for computing relative centrifugal force. A straight line drawn from the centrifuge speed to the rotor radius intersects the relative centrifugal force.

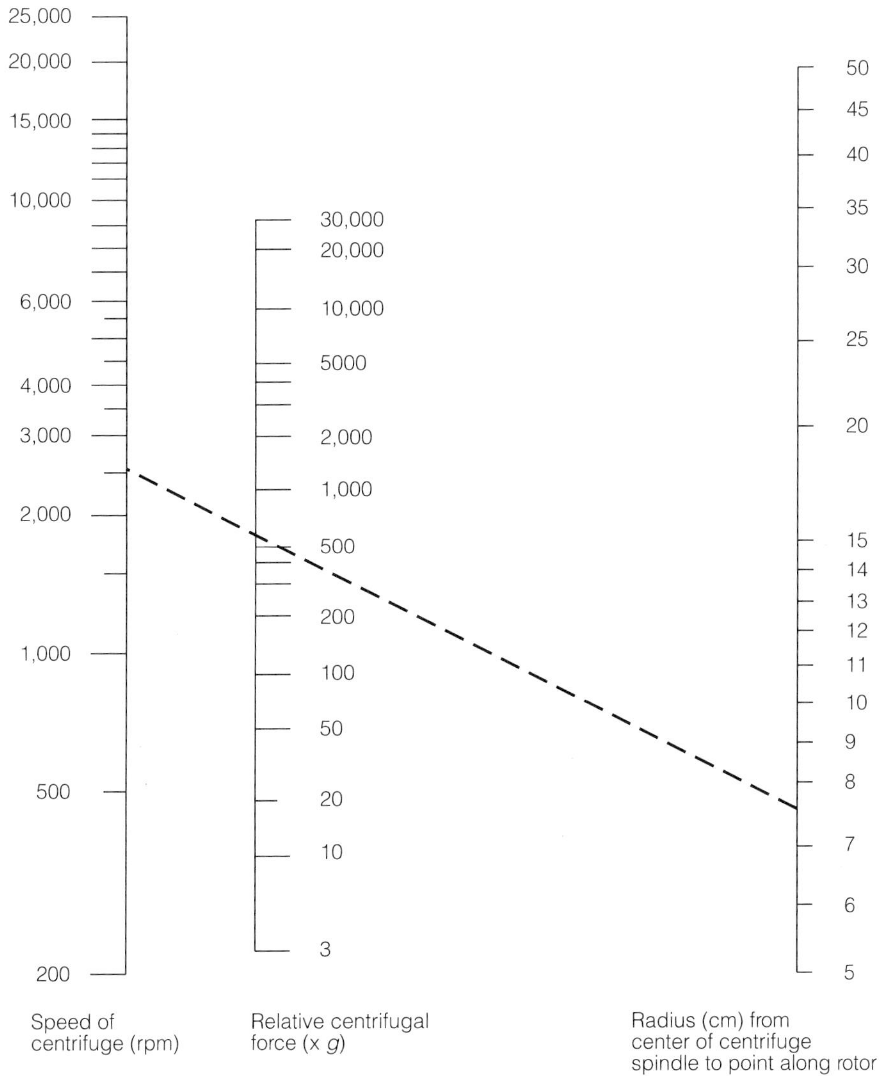

APPENDIX 7.2. Manufacturers of centrifuges.

Manufacturer	Type
Alfa Laval Inc. 2115 Linwood Ave. For Lee, NJ 07024 201–592–7800	Flow through Disk type
Beckman Instruments, Inc. 2500 Harbor Blvd. Fullerton, CA 92634 800–742–2345	Flow through ultra Laboratory
Electro–Nucleonics, Inc. 350 Passaic Avenue Fairfield, NJ 07006 201–227–6700	Flow through ultra
Sharples, Inc. 955 Merns Road Warminster, PA 18974 215–443–4234	Flow through Cylinder type
Westfalia Separator AG Postfach 3720 D–4740 Oelde Federal Republic of Germany (0 25 22) 77–1	Flow through Disk type Rotary filters

CHAPTER 8

PRECIPITATION AND EXTRACTION

Proteins differ in solubility, and the difference provides a basis for separation. The solubility of a protein in various solvents is largely due to the distribution of hydrophilic and hydrophobic residues on the surface of the molecule. Although hydrophobic groups tend to be sequestered in the interior of the three–dimensional structure of the protein, selected pockets or patches of hydrophobic residues will be present on its surface. These surface residues interact with ions, other proteins, and solvent molecules in ways that can be controlled to produce transfer to a different phase, either to a solid phase (precipitation) or to a liquid phase (extraction).

The degree of purification that can be obtained from a single–stage precipitation or extraction is small compared to that from such high–resolution steps as chromatography. For this reason extraction and precipitation are most commonly employed in the early part of the process, where their ability to reduce the mass or volume significantly is most valued, and they are operated under conditions where the amount of product lost and not recovered is minimized. Multistage operations are possible, and in some cases may give high–resolution separations comparable to those of column chromatography.

Precipitation occurs when the solubility of the protein in solution is reduced beyond some critical value. In aqueous solution this may occur by increasing the concentration of salt or some other material such as polyethylene glycol (PEG), that is, "salting out," or by decreasing the salt concentration, "salting in." Precipitation may also be accomplished by the addition of organic solvents.

Extraction is a process in which the protein of interest is transferred from the existing aqueous phase to another phase, either aqueous or organic. Aqueous partitioning exploits the varying distribution of proteins between two immiscible water–based phases. Conditions are usually such that biological activity is preserved. Two–phase aqueous systems made from PEG and dextran or salt are finding increased utility in large–scale protein purification. Extraction with reversed micelles and organic solvents is also used.

8.1 PRECIPITATION

Many precipitating agents are available, including salts, polyelectrolytes, nonionic polymers, heat, pH adjustment, and organic solvents. Several examples are given in Table 8.1. Glatz and Fisher (1986) have described precipitation as a multistep process of formation of submicron particles followed by aggregation of particles that are broken up by shear forces. Belter et al. (1988) describe precipitation as a six–step process: initial mixing, nucleation, diffusion–limited growth, flow influenced growth, flocculation, and separation by centrifugation. The rate–limiting step, which varies with the nature of the protein and the precipitating agent, must be determined experimentally. Scopes (1987) describes solubility in aqueous solution as the result of polar interactions between the protein and water molecules, ionic interactions between the protein and ions present, and repulsive electrostatic forces between like–charged molecules.

A change in the salt concentration of a solution may cause proteins to precipitate in either of two ways: As the salt concentration decreases, ionic interactions between protein molecules increase, leading to aggregation and precipitation by salting in; or as the salt concentration increases, the number of water molecules available for shielding hydrophobic groups decreases and binding of proteins to each other increases, leading to precipitation by salting out. (Remember them this way: In salting–*in*, addition of salt causes the protein to go *in*to solution; in salting–*out*, addition of salt causes the protein to fall *out* of solution.)

Many proteins precipitate at similar salt concentrations, and addition of salt to a crude solution results in a mixture of proteins in the solid phase. However, the range of concentration of salt in which a protein is soluble is often so narrow that all a particular protein may be removed from solution, resulting in substantial enrichment of the solid phase for the product. Scopes (1987) has found on the basis of empirical studies that the solubility of proteins follows the relation

$$\log \text{solubility} = A - m\,(\text{agent concentration})^{\mathrm{n}} \tag{8.1}$$

where A and m are constants dependent on temperature and pH, the agent in this case is salt, and n is equal to 1 for pure proteins or is greater than 1 for mixtures of proteins.

The efficiency of anions for precipitating proteins more or less follows the Hofmeister series (see Chapter 4):

$$\begin{aligned} &SCN^- > ClO_4^- > NO_3^- > Br^- > Cl^- > \text{acetate} \\ &\quad > \text{citrate} > HPO_4^{2-} > SO_4^{2-} > PO_4^{3-} \end{aligned} \tag{8.2}$$

Table 8.1. Agents used to precipitate proteins.

Agent class	Example	Mechanism
Polar solvent	Water	Salting-in (decrease in ionic strength)
Salt	Ammonium sulfate Potassium chloride	Salting-out (increase in ionic strength)
Polymer	Polyethylene glycol Polyethyleneimine	Salting-out
Temperature	Heat	Denaturation and aggregation
pH	Acid	Reduce ionic charge of protein
Non-polar solvent	Ethanol	Reduce activity of water

The more effective ions tend to be multivalent anions, such as sulfate, phosphate, and citrate. For cations Arakawa and Timasheff (1984) recommend monovalent ions and rank their effectiveness

$$NH_4^+ > K^+ > Na^+ \tag{8.3}$$

Common inexpensive salts include ammonium sulfate, potassium phosphate, sodium citrate above pH 8, and sodium sulfate at high temperatures. (Salts such as ammonium sulfate are frequently contaminated with metal ion impurities that may have a deleterious effect on proteins. The solubility of ammonium sulfate in water at 20°C is 533 g/l. A saturated solution of ammonium sulfate is about 4 M and has a density of 1.23 g/cm^3). Inclusion of EDTA will sequester these impurities. Divalent metal ions, however, have also been used to precipitate proteins. Zaworski and Gill (1988), for example, have shown that the addition of millimolar amounts of zinc chloride precipitated several recombinant proteins, with recovery effected by resuspension in buffer containing EDTA. The solubility is sensitive to pH; careful control gave recovery yields of 80 to 100 percent of activity.

The precipitation of proteins by addition of salt is common in laboratory practice but is often difficult to reproduce on a large scale. The concentration of saturated salt solutions varies with temperature, requiring strict control to avoid lot–to–lot variation. Mixing of large volumes is more difficult than that of small volumes, and local variations in salt concentration may occur. Such variations may lead to precipitation of more (and different) proteins at different locations within the tank. The density of precipitated protein is often similar to the density of concentrated salt solution, making separation by centrifugation difficult. A high concentration of salt in the product may be helpful if the next step is hydrophobic interaction chromatography but may need to be reduced for ion exchange chromatography. Although natural proteins are usually soluble in salt solutions of physiological concentration (0.15 to 0.2 M), recombinant proteins do not fit any general solubility classifications, some being quite different from proteins found in nature as a result of fusion or other manipulations.

Jendrisak (1987) has used polyethyleneimine (PEI), a positively charged molecule with a pK_a of 10 to 11, to precipitate nucleic acids and proteins selectively. Proteins thus precipitated may be solubilized by buffers of slightly higher ionic strength.

The nonionic polymer polyethylene glycol (PEG) precipitates proteins in aqueous solution without denaturing the proteins or otherwise interacting with them. PEG solubility is fairly insensitive to temperature, obviating the need for strict temperature control. Ingham has found that the solubility of proteins in PEG follows Equation 8.1, with n equal to 1 over a wide range of protein concentration. PEG is available in a wide range of molecular weights, with the range 4,000 to 6,000 D appearing frequently in the literature. Larger molecular weights add viscosity without improving the separation, though smaller sizes may give better resolution. Inclusion of a buffer compound is necessary to counteract the intrinsic buffering capacity of PEG. According to Ingham (1984) proteins typically precipitate in the PEG concentration range 3 to 30 percent (weight per volume). PEG may be removed from protein solutions by several methods, including adsorption of the protein on ion exchange resins, salt induced phase separation (see below), and ultrafiltration.

Temperature changes effect precipitation in two ways: Most proteins and many salts exhibit decreasing solubility with decreasing temperature; at high temperatures (for example, near boiling) proteins become denatured, lose their tertiary and secondary structure, and expose hydrophobic groups that bind with other molecules and form insoluble aggregates.

Proteins near their isoelectric point have little charge with which to repel other proteins and are therefore subject to hydrophobic attractive interactions that lead to aggregation and precipitation. Thus a change in the pH of the solution may be sufficient to cause precipitation of the solution. Proteins soluble at pH 8 may be insoluble at pH 6 and vice versa. The decreased solubility of a protein at its isoelectric point is particularly helpful in optimizing precipitation by salting out.

Organic solvents have been employed extensively for the precipitation of proteins. The Cohn process, used at very large scales for separation of blood–plasma fractions, relies on the variable solubility of proteins in ethanol at different temperatures. The addition of organic solvents, such as ethanol or acetone, reduces the water activity, with the solubility of proteins decreasing with increasing organic concentration. Highly hydrophobic proteins may be soluble in pure organic solvents, but most proteins will precipitate. Scopes (1987) has found that, other things being equal (including isoelectric point and hydrophobicity), the larger the molecule, the less organic solvent needed for precipitation to occur.

Organic solvents must be water soluble and in addition to ethanol and acetone include methanol, propanol, and other alcohols and ketones. Safety considerations eliminate such alternatives as ether. The opportunity for denaturation is high but can be reduced by operating at subzero temperatures. Acetone is less likely to cause denaturation than alcohol, and shorter–chain alcohols are less disruptive to proteins than are longer–chain alcohols. High salt concentrations also contribute to denaturation. Scopes (1987) has found that most enzymes precipitate in the range of 20 to 50 percent organic solvent and that above 50 percent solvent only proteins smaller than 15,000 D will remain in solution.

Most precipitates are mixtures of many different proteins, and if formed early in the process, may contain cell membranes, nucleic acid, and various particles. Precipitation may be carried out in two steps: the first precipitation step to remove proteins less soluble than the product, the second to precipitate the product. Belter et al. (1988) suggest that, when a precipitation step is scaled up, the kinetics should be held constant. For example, if in the small scale a precipitate of 500 μm in diameter forms in 5 minutes, the same kinetics should be sought at the large scale. They suggest that this can be accomplished if the power per unit volume in the mixer is held constant.

Often a crude fractionation can be obtained through adsorption onto a solid. These substances are usually used in a batch process by addition to the aqueous solution, with the resulting flocs removed by centrifugation or filtration. Less frequently the product protein may be adsorbed and then desorbed following the separation of the solids from the liquid. The solids with the adsorbed material are commonly disposed of in landfills, which limit their use to small–volume operations.

Materials used for batch adsorption include cross–linked polymeric particles of 0.1 μm diameter or larger with weakly or strongly charged anionic and cationic groups, fumed silica, and diatomaceous earth.

8.2 LIQUID–LIQUID EXTRACTION

8.2.1 Aqueous Partitioning

When a salt is added to an aqueous solution of a polymer such as polyethylene glycol (PEG), two immiscible phases are formed. The top, or less–dense phase, is usually enriched with polymer, and the lower, denser phase is enriched with salt. The distribution of any substance between these two phases can be described by a partition coefficient K:

$$K = \frac{c_t}{c_b} \tag{8.4}$$

where c_t is the concentration of the substance in the top phase and c_b is the concentration of the substance in the bottom phase. The partition coefficient K is equal to unity when the concentration of the substance is equal in both phases. A partition coefficient of 1,000 means that the concentration of the substance in the top phase is 1,000 times greater than its concentration in the bottom phase. The further the partition coefficient is from 1, the greater will be the recovery.

The distribution of a particular substance, such as a product protein, between the two phases is dependent upon the nature of the protein and the nature of the two phases. For example, a more ionic protein would be found in the high–salt phase, while a hydrophobic protein would be found in the polymer phase. Several parameters can be adjusted to alter the distribution of a particular protein, including the chemical nature of the primary polymer, the chemical nature of a secondary polymer, the molecular weight of the primary and secondary polymers, the type of salt, the concentration of polymers and salts, the concentration of protein added, the pH, the ionic strength, and the temperature. The large number of variables requires extensive experimentation for the optimizing of the process, but it also allows a wide range of conditions and makes the method applicable to a large number of proteins. Partitioning is independent of product concentration over a wide range. For such large particles as cells one–sided partitioning is common; cells, cellular material, and precipitated proteins are often found at the interface between the two phases. Dove and Mitra (1986) used a mixture of 20 percent (weight–per–volume) PEG 3350 at various salt concentrations to optimize the recovery of several proteins, including immunoglobins, albumin, and alpha–1 antitrypsin from Cohn–fractionated plasma.

The value of the partition coefficient can be determined more easily by experimental means than from fundamental thermodynamic data. In systems of PEG and salt as well as systems of PEG and dextran, the PEG–rich phase has lower density and forms the top layer. Many proteins follow the theory of Flory Huggins, which predicts that lowering the molecular weight of PEG will increase the partition coefficient, though Hummel et al. (1984) have found some exceptions. Similarly, Kula (1979) has shown that increasing the molecular weight of dextran also increases the partition coefficient, but to a lesser extent than does changing the PEG weight.

The distribution of proteins between the phases is greatly influenced by pH. Highly ionized proteins will favor the lower phase, while more hydrophobic proteins will favor the upper PEG phase. In addition, Kula (1987) has found that in PEG–dextran systems, phosphate ions will preferentially remain in the dextran phase, causing the displacement of negatively charged proteins to the PEG phase. Kula further suggests that, because the partition coefficient varies only weakly with temperature and the phase–forming polymers tend to stabilize proteins, large–scale operations can be safely conducted at room temperature.

Both PEG and dextran are available commercially in several molecular–weight fractions. The volume of the polymer phase relative to the volume of the salt phase tends to decrease with increasing salt concentration and increasing pH. Because both phases are aqueous, equilibrium is achieved quickly with low power requirements. The time required to separate the two phases under unit gravity depends on the protein load. Kula (1985) has shown that with high loads a stationary interface is reached after 15 to 20 hours; when no solids are present, a stationary interface is reached in less than 2 hours. Centrifugal separation will naturally speed up the process.

The composition of each phase can be determined for the total system composition from the phase diagram. Phase diagrams were first described for aqueous two–phase systems by Albertsson (1958). Figure 8.1 shows phase diagrams for the systems polyethylene glycol–potassium phosphate and polyethylene glycol–dextran. The curve represents the division between the two immiscible phases. On the convex side of the curve (to the left and below) only one phase exists. Above and to the right of the curve two immiscible phases exist. The composition of two phases in equilibrium with each other is given by the tie line. The end points of a tie line that passes through the system composition give the respective compositions of each of the two phases. For example, in Figure 8.1, a system with the total composition of 5 percent polyethylene glycol and 5 percent dextran (point *A*) will divide into two phases, one rich in polyethylene glycol (7 percent polyethylene glycol, 1 percent dextran—point *B*) and the other rich in dextran (12 percent dextran, 2 percent polyethylene glycol—point *C*). A change in the total system composition to a longer tie line generally gives a higher partition coefficient for the protein at the expense of higher concentrations of phase–forming polymers and therefore higher costs.

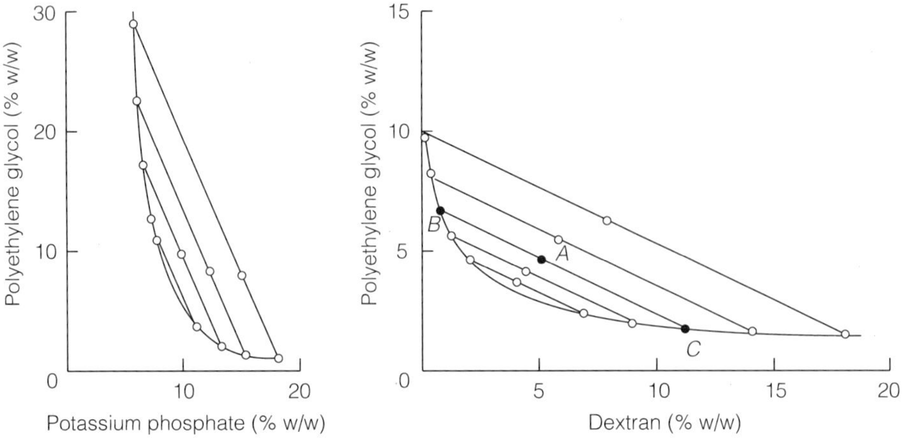

Figure 8.1. Phase diagrams for aqueous two-phase systems of polyethylene glycol–potassium phosphate and polyethylene glycol–dextran. From Albertsson (1986); used by permission.

One drawback of two–phase partitioning is that residual PEG must be removed from the product in a later step. This may be done by initial partitioning of the protein into the PEG phase in a PEG–dextran system, followed by partitioning into a salt phase in a PEG–salt system. Low–molecular–weight polymers may be removed by diafiltration. Ion exchange and hydrophobic interaction chromatography may also be used, as the phase–forming polymers do not usually bind to these. Gel filtration chromatography is also applicable. Partitioning polymers are expensive; so the more protein that can be loaded, the lower the cost of materials (on a per–unit–of–product basis). The advantages of two–phase partitioning include high recovery rates, low capital costs, and simple operation. The partition coefficient is independent of volume and therefore can be maintained during scale up. Aqueous partitioning may also serve to concentrate the product. Affinity linkages on the polymer have also been proposed.

8.2.2 Reversed Micelles

Reversed micelles are a two–phase system consisting of droplets of aqueous phase separated from an organic phase by a surfactant layer on the surface of the droplet. They are formed by vigorous agitation of the aqueous solution with surfactant and solvent. The surfactant forms a layer with its polar head inside the micelle and nonpolar tail outside the micelle, as shown in Figure 8.2. Proteins in the aqueous solution can be made to move into or out of the micelles with changes in the pH, ionic strength, and surfactant concentration. The micelles can be recovered by centrifugation or filtration.

Goklen and Hatton (1985) have demonstrated that transfer of proteins from the organic phase to the interior of the reversed micelle occurs by electrostatic attraction when the charge on the protein is opposite to that of the surfactant head. If the surfactant head is negatively charged, the transfer of protein into the reversed micelle occurs only below the pI of the protein. Thus pH plays a significant role in solute partitioning.

Hydrophobic interactions also play a role in protein solubility within the micelle: A high ionic strength within the aqueous phase will exclude hydrophobic proteins from the micelle. Woll et al. (1987) found that at low protein concentrations (below 8 mg/ml) the partitioning of protein between phases was independent of protein concentration. Woll also found that the capacity of the organic phase for protein from a given feed stream was limited by the amount of surfactant available to form micelles to contain the aqueous phase.

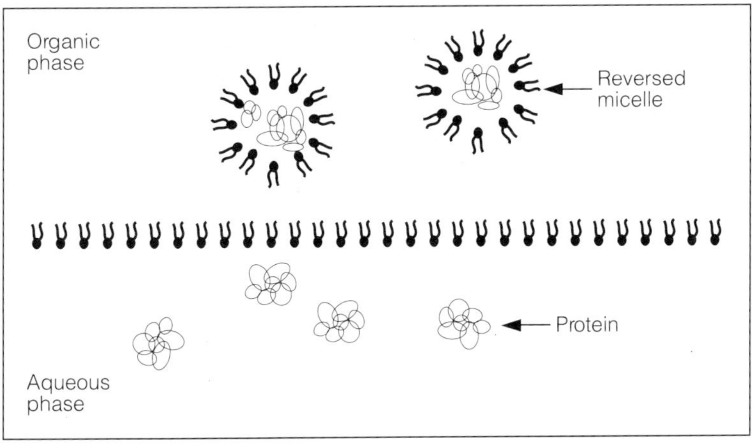

Figure 8.2. Diagram of partitioning of proteins into micelles in organic phase.

The use of reversed micelles is limited by the cost of solvent and surfactant, including their possible regeneration and reuse. As in other two–phase systems the separation of the phases and recovery of the protein must be taken into account. The use of specific affinity surfactants may also be possible.

8.2.3 Other Partitioning Systems

Lovrien et al. (1987) describe a system utilizing three phases formed by addition of ammonium sulfate and *t*–butanol to water. An intermediate layer is formed between the upper solvent and lower water into which proteins may be partitioned.

Supercritical fluid extraction has been used to recover small proteins and peptides from mixtures. Large molecules are not well solubilized under supercritical conditions.

For systems with partition coefficients near 1, multistage units will increase the resolving power, although at increased cost. The addition of centrifugal force to separate the phases, as in the continuous–flow separator developed by Sanki Engineering (Japan), allows high–resolution separation in liquid–liquid systems. The Sanki separator provides high efficiencies through many countercurrent contacts, making liquid–liquid chromatography possible.

The surface properties of proteins and cells that attract them to flocculating agents, such as surfactants, can serve as a basis for separation. Cells are kept apart by electrostatic forces, hydration forces, and stearic hindrance of protrusions. Changes in these can result in aggregation of cells; aggregated cells may be easier to separate than are nonaggregated cells, for example, by centrifugation. Warne and Bowden (1987) suggest that flocculation can be accomplished by addition of high–molecular–weight polymers, such as polyacrylamide, which serve to form bridges between cells.

REFERENCES

Albertsson, P. A., Partition of Cell Particles and Macromolecules, 3rd ed., John Wiley & Sons, New York (1986).

Albertsson, P. A., "Particle fractionation in liquid–liquid two–phase systems," *Biochimica et Biophysica Acta* 27:378–395 (1958).

Arakawa, T., and S. N. Timasheff, "Mechanism of protein salting in and slating out by divalent cation salts: balance between hydration and salt binding," *Biochemistry* 23:5912–5923 (1984).

Belter, P. A., E. L. Cussler, and W. S. Hu, Bioseparations: Downstream Processing for Biotechnology, John Wiley & Sons, New York (1988).

Dove, G. B., and G. Mitra, "Recovery of proteins from polyethylene glycol–water solution by salt partition," in Separation, Recovery, and Purification in Biotechnology, J. A. Asenjo and J. Hong (eds.), *ACS Symposium Series 314*, American Chemical Society, Washington (1986).

Glatz, C. E., and R. R. Fisher, "Modeling of precipitation phenomena in protein recovery," in Separation, Recovery, and Purification in Biotechnology, J. A. Asenjo and J. Hong (eds.), *ACS Symposium Series 314*, American Chemical Society, Washington (1986).

Goklen, K. E., and T. A. Hatton, "Protein extraction using reversed micelles," *Biotechnology Progress* 1:1 (1985).

Hummel, W., H. Schutte, and M. R. Kula, "New enzymes for the synthesis of chiral compounds", *Enzyme Engineering 7, Annals of the New York Acadamy of Science* 434:87–90 (1984).

Ingham, K. C., "Protein precipitation with polyethylene glycol," *Methods in Enzymology* 104:351–356 (1984).

Jendrisak, J., "The use of polyethyleneimine in protein purification," in Protein Purification: Micro to Macro, R. Burgess (ed.), Alan R. Liss, New York (1987).

Kula, M. R., "Extraction and purification of enzymes using aqueous two–phase systems," *Applied Biochemistry and Bioengineering* 2:71–95 (1979).

Kula, M. R., "Liquid–liquid extraction of biopolymers," in Comprehensive Biotechnology, A. Humphrey and C. L. Cooney (eds.), Pergamon Press, New York (1985).

Kula, M.R., "Use of phase partitioning to scaleup protein purification," in Protein Purification: Micro to Macro, R. Burgess (ed.), Alan R. Liss, New York (1987).

Lovrien, R., G. Goldensoph, P. C. Anderson, and B. Odegaard, "Three phase partitioning via t–butanol," in Protein Purification: Micro to Macro, R. Burgess (ed.), Alan R. Liss, New York (1987).

Scopes, R. K., Protein Purification: Principles and Practice, 2nd ed., Springer Verlag, New York (1987).

Warne, S. R., and C. P. Bowden, "Biosurface properties and their significance to primary separation: a genetic engineering approach to the utilization of bacterial auto–aggregation," in Separations for Biotechnology, M. S. Verrall and M. J. Hudson (eds.), John Wiley & Sons, New York (1987).

Woll, J. M., A. S. Dillon, R. S. Rahaman, and T. A. Hatton, "Protein separations using reversed micelles," in Protein Purification: Micro to Macro, R. Burgess (ed.), Alan R. Liss, New York (1987).

Zaworski, P. G., and G. S. Gill, "Precipitation and recovery of proteins from culture supernatants using zinc," *Analytical Biochemistry* 173:440–444 (1988).

FURTHER READING

Abbott, N. L., T. A. Hatton, "Liquid–liquid extraction for protein separations," *Chemical Engineering Progress* 84(8):31–41 (1988).

Walter, H., D. Brooks, and D. Fisher, ed., Partitioning in Aqueous Two–phase Systems: Theory, Methods, Uses and Applications to Biotechnology, Academic Press, Orlando (1985).

CHAPTER 9

FILTRATION

Filtration is widely used in biotechnology for the separation of substances on the basis of relative size. It is useful in both the removal of small particles from solution and the concentration and separation of soluble molecules.

By conventional usage *filtration* is defined as the separation of solid particles from liquid or gaseous streams. With membranes this may be extended to include the separation of dissolved solutes in solution. The primary role of the filter is that of a selective barrier through which certain parts of the process stream (the *filtrate*, or *permeate*) may pass but behind which other parts of the process stream (the *retentate*) are retained. The process can be so designed that the product of interest is found in either the filtrate or the retentate.

Filters may be classified by several features, such as the nature of the filter medium, the structure of the filter (including its morphological characteristics), the application, and the mechanism of filter action, such as adsorption, diffusion, and selective permeability (*perm–selection*). Although an increase in applied pressure may increase the rate of transport in microfiltration and ultrafiltration, it is the nature of the filter itself that controls which components transverse the filter and which are retained.

There are two types of membrane process that are most widely applicable in large–scale protein purification: microfiltration and ultrafiltration. These two types of filtration differ ostensibly by the size of the particles they treat; however, the mechanisms behind these processes also differ. Microfiltration is generally concerned with particles whose size ranges from about 0.05 μm to about 2 μm. Ultrafiltration overlaps this range, dealing with particles or molecules of approximately 0.2 μm down to approximately 200 nm. Figure 9.1 shows classifications of various particles, including those amenable to microfiltration and ultrafiltration. Other membrane processes having application in biotechnology include perm–selective membranes, electrodialysis, and reverse osmosis.

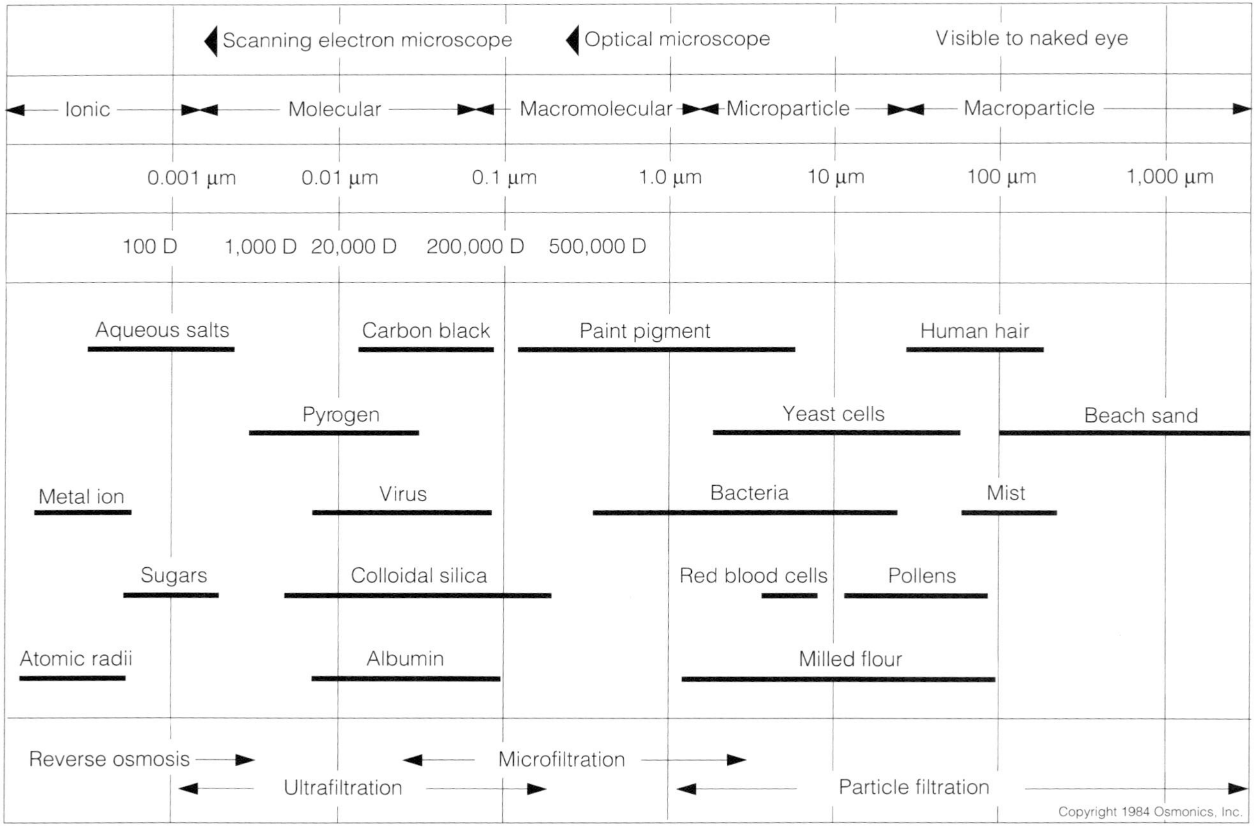

Figure 9.1. Classification of particles by size and applicable separation process. (Copyright 1984 Osmonics, Inc.; courtesy of Osmonics, Inc., Minnetonka, MN.)

Microfiltration is commonly used to remove suspended particles from a process fluid and comprises operations such as the recovery of cells from fermentation broth and the clarification of lysed–cell slurries. In this sense it is a unit operation that is comparable to, and offers the process designer an alternative to, centrifugation. Both unit operations selectively remove particles from aqueous mixtures, albeit by different mechanisms. Centrifugation relies on the sedimentation of the particles, that is, a difference in the densities of the particle and the medium. Membrane microfiltration, on the other hand, relies on retention of the particles behind a membrane screen with passage of the solution through the membrane. In practice both membrane microfiltration and centrifugation can be used to separate particles on the basis of size and are to some extent interchangeable.

Ultrafiltration is an effective technique for concentrating or separating dissolved molecules of different sizes. In ultrafiltration the pores of the membrane are small enough to prohibit the passage of proteins through the membrane while not prohibiting the passage of salt ions, buffers, and water. Therefore, the total concentration of the protein in solution can be raised by passing solvent through the membrane while retaining the protein. Another application of ultrafiltration membranes is diafiltration, which is an exchange of buffers similar to dialysis. Unlike dialysis, however, with ultrafiltration membranes this exchange can occur very quickly and easily at large volumes.

Ultrafiltration has several applications in biotechnology: It may be used during the initial separation of the cells from the fermentation or culture medium; it may be used after the cells are broken to separate the cellular fragments from the medium; it may be used to fractionate the protein solution, that is, either to have the product pass through the membrane with larger materials retained or else to have the smaller materials pass through the membrane with the product retained—or both in a bracketed operation. Ultrafiltration has been used successfully with shear–sensitive products, including labile viruses, as described by Trudel et al. (1983). One of the advantages of ultrafiltration is that it is easily scaled up to handle very large volumes. An advantage of ultrafiltration over other types of dewatering or concentration processes is that the material in solution—namely, the protein—does not change phase. For example, in dewatering by crystallization or organic precipitation, the protein undergoes a phase transition, which may result in denaturation of the protein. Ultrafiltration avoids this phase change and thus is a much more mild treatment of the protein. Another advantage of ultrafiltration is that small molecules pass readily through the membrane, and therefore the environment in which the protein finds itself with respect to buffers, salts, and so on, remains unchanged; that is, there are no changes in pH or ionic strength. One disadvantage of ultrafiltration is that it cannot take the product to dryness, as ultrafiltration systems are limited in the extent of the solids that they can handle.

Microfiltration and ultrafiltration will be discussed individually, with a description of the mechanism of each, the types of membrane used, the applications in biotechnology, and the scale up of these unit operations given in detail.

9.1 MICROFILTRATION

Microfiltration is the physical retention of particles behind a filter medium with passage of the process fluid through the filter medium. The particles are retained primarily because they are too large to pass through the openings in the filter medium although surface charge and other properties may be employed to augment retention.

Microfiltration media are of two types: depth filters and screen filters. This is a classification based on the morphology, or structure, of the filter. Depth filters consist of fibers or fibrous materials packed together, which trap particles as they travel through the tortuous paths between the fibers. As particles build up in the bed of the depth filter, they in turn make the pathways more tortuous and lead to further particle capture. Screen, or membrane, filters, on the other hand, are typically made from a membrane sheet having pores of very precisely controlled size. Like sieves, membrane filters separate particles by retaining them on the surface. Figure 9.2 shows a schematic representation of the two types of filter.

Screen, or membrane, filters can be further classified as either symmetric or asymmetric according to their substructure. Asymmetric membranes are also referred to as "skin membranes." Their basic structure is that of a sponge, with large pores topped by a very thin layer of definite, limiting, pore size.

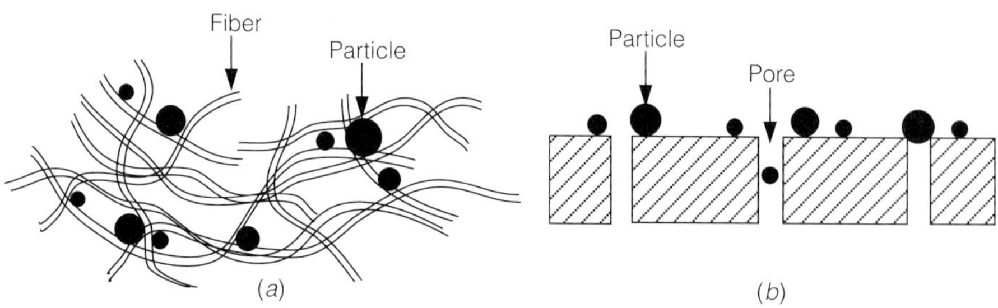

Figure 9.2. Schematic diagram of microfilters: (*a*) depth, (*b*) screen, or membrane.

The sponge offers mechanical support to the thin skin, which serves as the final barrier limiting the size of the particles that can pass the membrane. The terms "isotropic" and "anisotropic" are sometimes substituted for symmetric and asymmetric. Some membrane filters may also have tortuous paths. Particularly, isotropic membranes may have pores slightly larger than the smallest particle that they retain, but the tortuosity of the pore prevents particles from migrating through the filter. Figure 9.3 shows scanning electron micrographs of a cross section of an asymmetric membrane and of the surface of an isotropic membrane.

Filters are rated in terms of the size of particles that they retain. Depth filters cannot be rated absolutely in terms of minimum particle size because of the nature of the tortuous paths between the fibers. In a depth filter a particle of fairly large size can eventually wander through, given enough time and the right conditions in the fiber bed. On the other hand, a screen–type filter has a definite pore size, which restricts the size of the particles that can pass through it, and thus has a very definite particle–size cutoff.

Microporous membranes are designed to retain all particles above the rated size. That is, a 0.45–μm membrane is rated to retain all particles larger than 0.45 μm. Of course, particles that are approximately the same size may lodge partly in the pores and block them. Particles that are nonspherical in shape, such as rod–shaped particles, having a diameter smaller than 0.45 μm, may actually pass through these membranes. This is the case with some bacteria and the reason why 0.2–μm filters are commonly used for sterilization. Another common practice is the use of multiple filters, such as a 0.45–μm followed by a 0.2– or 0.1–μm filter. (Dimensions of some common organisms and cells have been listed in Table 6.1.)

The driving force for separation in microfiltration is the pressure drop across the filter. This pressure differential may be generated by pumping the fluid at a constant flow rate or by pressurizing the feed tank at a constant pressure. As the filter collects debris, the available pore area will diminish, increasing the pressure in a constant–flow system or decreasing the flow in a constant–pressure system. The rate of filtration is inversely proportional to the viscosity, and the viscosity is inversely proportional to the temperature. Lowering the temperature at a constant pressure will reduce the flow rate.

9.1.1 Depth Filters

Depth filters have the advantage of very large particle–holding capacity compared to screen filters of equivalent surface area. Depth media are also relatively inexpensive and so are often preferred for the treatment of very dirty streams. Depth filters have the disadvantage of a tendency to shed fibers; that is, small fibers come loose from the bed and migrate out into the finished product. For this reason depth filtration may be followed by screen filtration to remove any filter particles dislodged in the process and to ensure that all particles above a minimum size are removed. Depth filters or coarse screen filters are often used as prefilters upstream of a final clarifying filter, especially in the filtration of deformable particles. In addition to a strict particle–size cutoff and the absence of fiber shedding, another advantage of screen filters is that the material retained behind the screen can be easily recovered, whereas in the depth filter it is often embedded in the matrix of the filter.

Depth filters come in a variety of configurations. The pleated cartridge, which is very popular in the pharmaceutical industry, is used in a stainless–steel housing that can be steam sterilized. Additional surface area (dirt–holding capacity) can be obtained by increasing the length of the cartridge and by inserting multiple cartridges into a single housing. The stacked–disk arrangement offers good capacity in minimal volume and is useful with media that cannot be formed easily into pleated cartridges. Flat sheets and disks tend to have a lower strength than do pleated cartridges, in terms of the maximum pressure they can operate under. Plate and frame devices incorporating flat sheets can be built to virtually any capacity, with examples in the food industry including units as large as fifty single–sheet sections each of 1.5 m^2, or 75 m^2 (800 ft^2) total area.

Depth filters are constructed of fibers, fibrous material, particles, or some combination of these. Fibers include cellulose in the form of paper, fabric, or compacted bundles; fiberglass; and nylon. Particles include sintered steel, alumina, and other ceramics. Filtration with large particles, such as sand or diatomaceous earth, is used in water purification but is outside the scope of this chapter.

9.1.2 Membrane Microfilters

The controlled pore size of membrane filters makes them more attractive than depth filters for many operations. While depth filters are always operated in the "dead–end" mode, with the process stream's flowing normal to the plane of the media, screen filters may be operated in the "cross–flow" mode, the process stream flowing parallel to the plane of the media. In the dead–end mode the fluid butts up against the filter, and the particles collect against the media. In the cross–flow mode the fluid is swept across the surface of the filter and the particles are kept in suspension above the media.

Comparison of the surfaces of the membranes shown in Figure 9.3 illustrates the effectiveness of these two schemes. In the case of the asymmetric membrane little capacity results from dead–end operation, while cross–flow operation works well. The isotropic membrane, on the other hand, has good capacity in the dead–end mode.

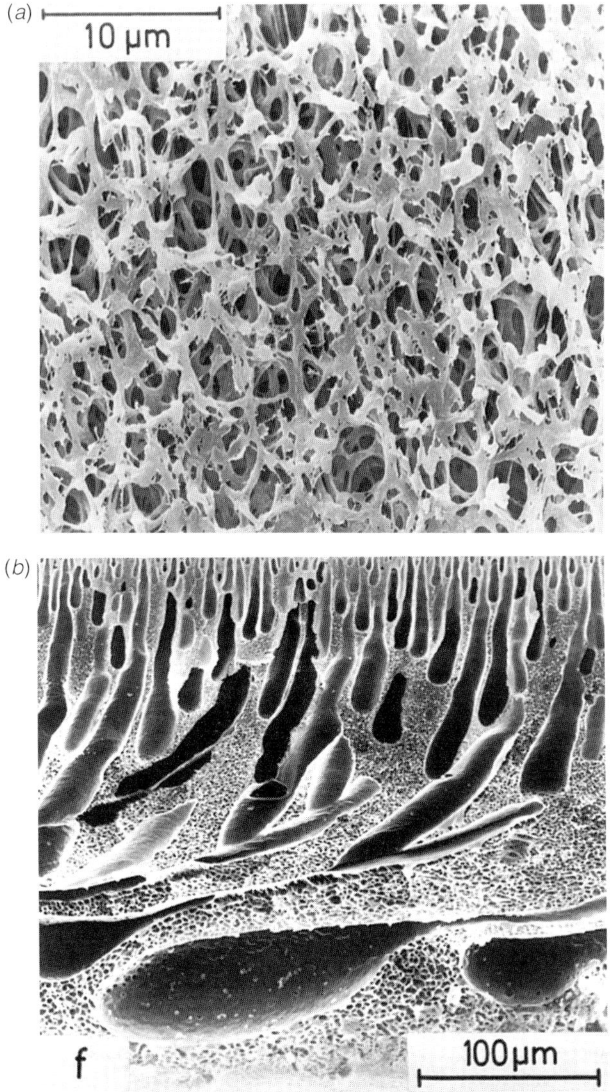

Figure 9.3. Photomicrographs of (*a*) cross section of asymmetric ultrafiltration membrane and (*b*) surface of isotropic microfiltration membrane. From Pusch and Walch (1982) and Pusch (1987); used by permission.

Tangential–flow microfiltration differs from tangential–flow ultrafiltration in more than just the membrane pore size. Tangential–flow microfiltration is restricted to the separation of particles; ultrafiltration can also separate soluble species. Microfiltration membranes are usually isotropic (symmetric); ultrafiltration membranes are usually anisotropic (asymmetric). The rated pore diameter (RPD) is a factor used to describe microporous membrane filters and is typically 0.1 to 10 μm.

Cross–flow filtration with isotropic membranes is subject to plugging of the membranes by deformable particles, which may be trapped by the uneven surface and held against the pore openings. The cross–flow motion of particles across membranes of asymmetric and isotropic types may be likened to the washing with a hose of dirt or debris across driveways of smooth concrete and rough gravel. In the case of the smooth concrete it is fairly simple. The cracks and crevices in the concrete are much smaller than the particles of debris which one wishes to wash off, and therefore the debris moves across the concrete surface quite readily. This is comparable to the asymmetric membrane. The isotropic membrane however is much more like the gravel drive in that the surface is less smooth, and therefore the holes may be more likely to become plugged by compaction of deformable materials.

Membranes for filtration are made of polymeric materials or ceramics. Among the considerations of filter–media selection is the tendency that some media have to bind proteins. This can result in a diminution of product yield. Cellulose acetates bind less protein than do mixed esters of cellulose, nitrocellulose, or nylon. The further the protein is from its isoelectric point, the more soluble it will be and the better it will pass through the filter. Olson (1983) has found that albumin does not bind to mixed esters of cellulose or nylon. Anisotropic membranes are typically more expensive to manufacture, and therefore their use is often limited to ultrafiltration on a cost basis rather than on pure functionality.

Membrane filters are also useful in sterilization of solutions by the removal of microorganisms. A 0.45– or 0.22–μm membrane is commonly used to sterile–filter a previously clarified solution. Filtration may be conducted at various temperatures; it is commonly carried out at room temperature as a matter of convenience, but it may be carried out at lower temperatures to reduce microbial growth or at higher temperatures to reduce the viscosity of the solution being handled.

9.2 ULTRAFILTRATION

As in microfiltration, ultrafiltration consists in the transport of solution through the filter medium or membrane in the ultrafiltrate and retention of the solute behind the membrane in the retentate. The driving force for the separation is the pressure difference across the membrane, also referred to as the "transmembrane pressure." The pores of an ultrafiltration membrane are sufficiently small to prohibit the passage of large molecules. The terms "large" and "small" are relative, and membranes are available with a nominal molecular–weight cutoff (NMWCO) from 5,000 up to 500,000 D or larger. In ultrafiltration the molecules that pass into the pores of the membrane are carried through with the process fluid, typically water.

Several factors affect the ability of ultrafiltration membranes to retain molecules. As listed by Cheryan (1986), these include the size and shape of the molecule, the type and configuration of the membrane material, the presence of other solutes, the concentration of the retained species, and the adsorption of solutes by the membrane.

9.2.1 Flux

The flux is defined as the rate of transport through the membrane, in units of mass or volume per cross–sectional area per unit time (for example, liters per square meter per hour). One may be concerned with the flux of water in a concentration step or with the flux of a particular species in a fractionation step. The flux of a species depends on the properties of the feed solution, while the retention of a species depends on the properties and nature of the membrane.

In ultrafiltration at the membrane surface a layer forms that is enriched for species that do not pass the membrane. The phenomenon of increased concentration between the bulk solution and the membrane surface is called "concentration polarization" or "gel polarization." The boundary layer that forms, called the "concentration–polarization layer," is shown schematically in Figure 9.4. Water from the bulk solution moves through this gel layer, carrying along small solutes that traverse the membrane under the driving force of the transmembrane pressure. Large solutes are also carried along but will not pass the membrane; this leads to the increased concentration found in the gel layer. Solutes that do not pass the membrane are returned to the bulk solution by diffusion away from the membrane. The flux is governed by the rate of transport through the gel layer and through the membrane. Except in rare cases of high membrane resistance (such as occurs with a membrane having very few pores) transport through the gel layer limits the flux. Back diffusion is the rate–limiting step that governs the formation of the gel–polarization layer.

There are three ways to reduce the concentration difference across the gel–polarization layer, as described by Bruin et al. (1980) and diagrammed

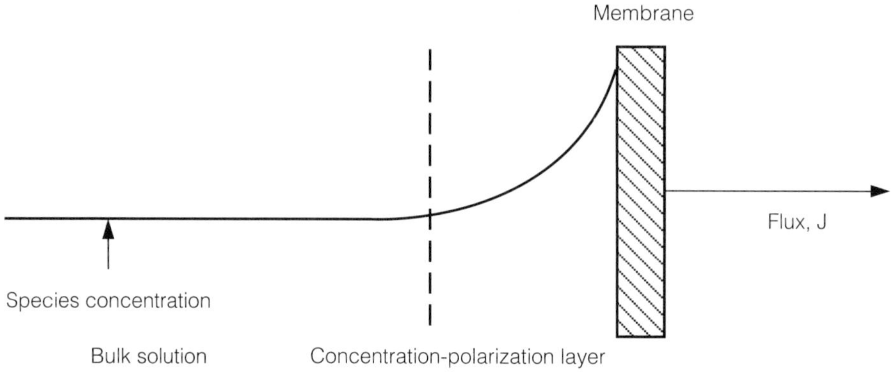

Figure 9.4. Schematic representation of concentration polarization.

in Figure 9.5: The first is to maintain a low transmembrane flux, either by a low transmembrane pressure differential or by a high membrane resistance. The second is to lower the concentration difference between the bulk solution and the membrane surface. The third is to reduce the concentration difference between the feed and the filtrate, by maintenance of a low concentration in the feed. The first alternative is impractical if a high flux is desired. The third is impractical in the case of ultrafiltration for concentration, where a low solids concentration in the feed cannot be maintained; but in the case of diafiltration a low solids concentration can be maintained by using a continuous method. In the case of concentration the concentration polarization can be prevented from increasing by maintaining a low concentration factor. A low transmembrane flux is counterproductive in the sense that the lowest–cost operations tend to be those that go the fastest; maintaining a low flux will lengthen the time of the process. Therefore the most productive means of reducing the concentration polarization is the second way, namely, lowering the concentration difference between the bulk and the membrane surface. All practical approaches for increasing the flux of ultrafiltration membranes address this point.

To lower the concentration difference between the bulk and the membrane surface, the rate of mass transfer through the boundary layer must be increased. There are two ways to do this: increase the diffusion rate or decrease the thickness of the layer. For the diffusion rate to be increased the process stream must be modified, such as by reduction of the viscosity or raising of the temperature to increase the diffusivity. The thickness of the boundary layer may be reduced by increase of the turbulence either tangential to the membrane or at right angles to the membrane. In the tangential direction either the liquid or the membrane may be moved. A short flow length will keep the flow rate high over the full membrane area. Static mixers placed within the flow channel may induce turbulence. Taylor

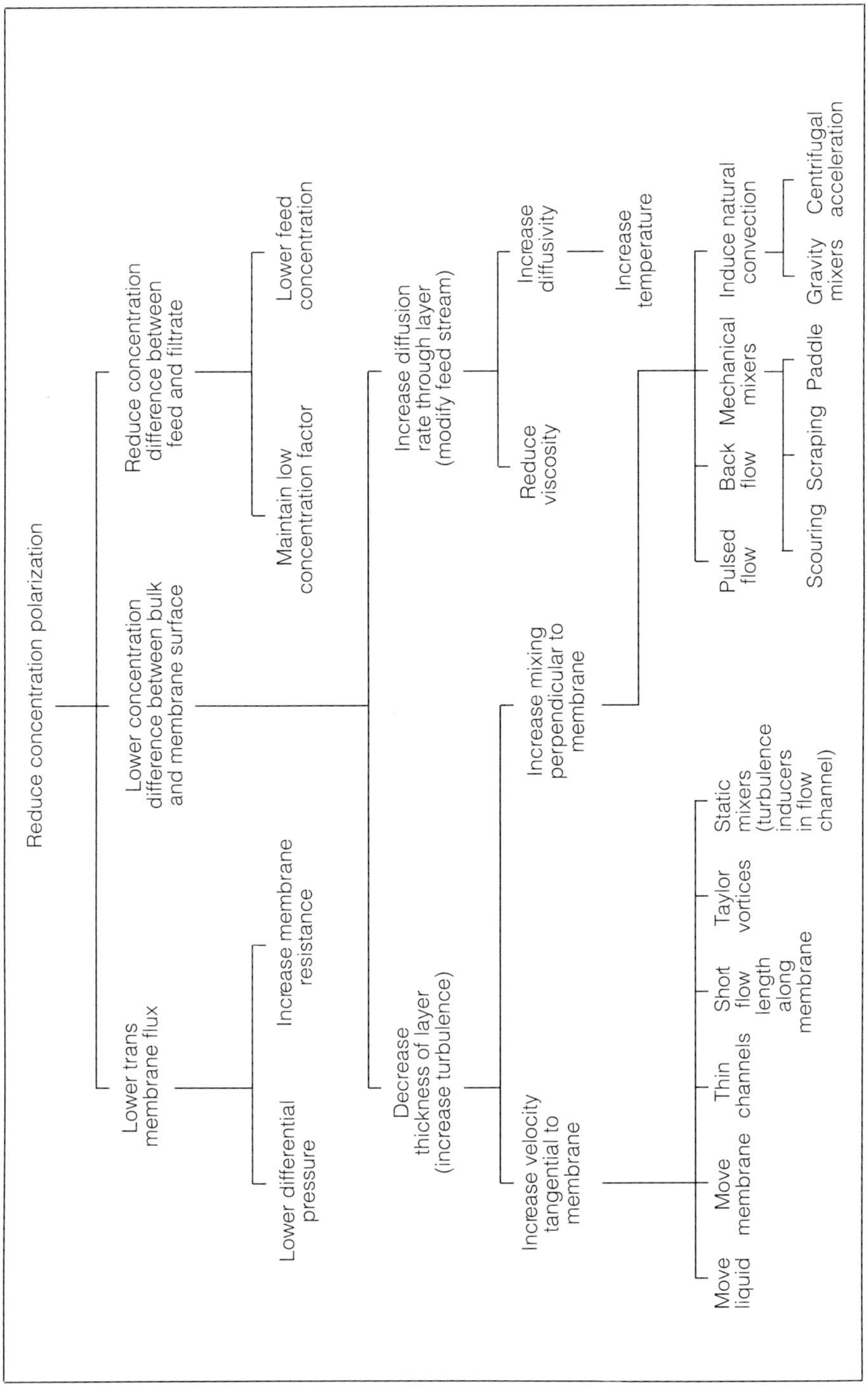

Figure 9.5. Alternative ways of reducing concentration polarization. After Bruin et al. (1980).

vortices may be induced by movement between the membrane and a nearby surface. In the perpendicular direction mechanical mixers may agitate the bulk and scrape or scour the membrane. Back flow through the membrane will disrupt the boundary layer and create mixing. Natural convection may be induced by gravity or centrifugal acceleration. One consideration, therefore, in looking at ultrafiltration equipment is to consider designs that minimize the resistance of the gel–polarized layer; practical ultrafiltration requires movement of the fluid perpendicular or parallel to the membrane surface.

The major operating parameters that affect the flux are pressure, temperature, feed concentration, and flow rate. In most models of membrane transport the flux is proportional to the pressure. A simple model based on the resistances of the membrane and the gel layer is

$$J = \frac{P}{\mu(R_m + R_g)} \tag{9.1}$$

where J is the flux, P is the transmembrane pressure, μ is the viscosity, R_m is the membrane resistance, and R_g is the gel–layer resistance. Formation of a gel–polarized layer at the surface of the membrane limits convective flow; transport of molecules through this layer depends on diffusion. Diffusion through this layer increases with increasing feed concentration and also with increasing temperature.

If the flow rate and the corresponding turbulence in a cross–flow channel is increased, then the thickness of the boundary layer will decrease. This leads to an increase in the rate of transport to the membrane and, consequently, an increase in the flux.

Physical properties of the liquid feed stream also have a bearing on the rate of transport through the membrane. Figure 9.6 shows the flux as a function of transmembrane pressure at different protein concentrations. As the protein concentration increases, the flux at a given pressure decreases. Figure 9.7 shows the flux of the filtrate as a function of protein concentration for various recirculation rates. As the cross–flow velocity increases, the flux increases for a given protein concentration. Concentration effects in the direction perpendicular to the membrane wall are not negligible. Most aqueous protein solutions are quite dilute, and the density is relatively close to that of water. Viscosity, however, may be different from that of water because of several effects and may have considerable influence on the behavior of the flow.

Aqueous solutions of proteins may be treated as aqueous solutions of polymers and have many of the same characteristics. In general the viscosity tends to decrease with increasing temperature and tends to increase with increasing feed concentration. For typical protein solutions the drop in filtrate rate with decreasing temperature is 1 percent per Fahrenheit degree. Flux also increases with increasing diffusion coefficient for both turbulent and laminar flow. The diffusion coefficient can be increased by increasing the temperature.

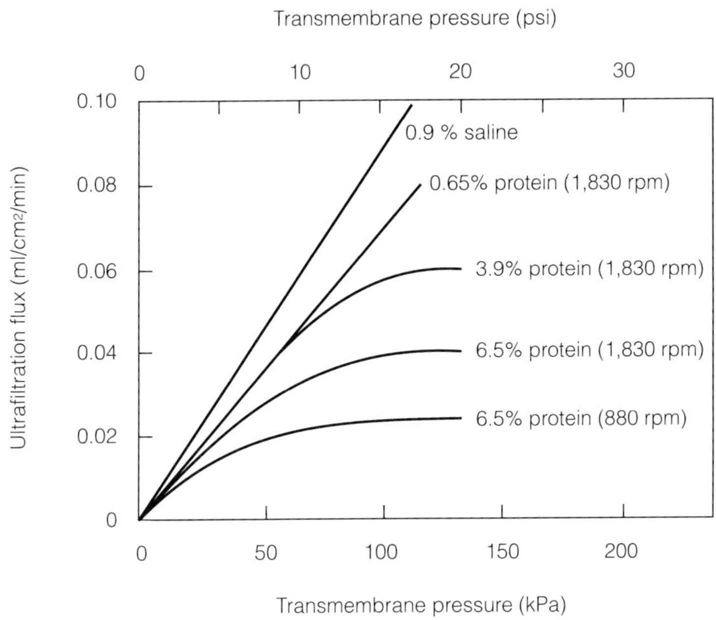

Figure 9.6. Flux as a function of transmembrane pressure at various concentrations of bovine serum albumin. [Reprinted by permission from M. C. Porter, *Ind. Eng. Chem., Prod. Res. Develop. II* 3:234–248 (1972). Copyright 1972 American Chemical Society.]

When the transmembrane pressure is below the gel–polarization–forming pressure, an increase in pressure generally results in an increase in flux, though not necessarily in a linear fashion. As the pressure increases above the gel–forming pressure, the gel–polarization layer forms and remains stable at that pressure, and transport through the gel polarization layer is driven by concentration and not by pressure. Therefore an increase in the pressure may increase flux only until the gel–polarization–layer thickness is increased and again becomes stable. This is illustrated in Figure 9.8, which schematically describes the increased gel–layer thickness with increased transmembrane pressure.

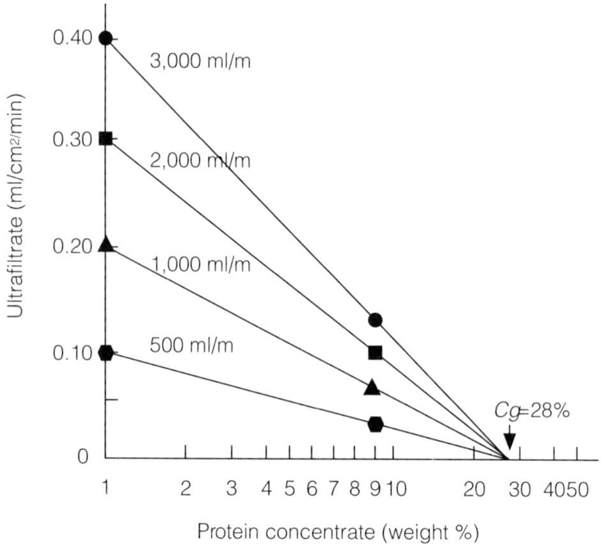

Figure 9.7. Flux as a function of protein concentration for various recirculation rates. *Cg* is the concentration in the gel layer. [Reprinted by permission from M. C. Porter, *Ind. Eng. Chem., Prod. Res. Develop. II* 3:234–248 (1972). Copyright 1972 American Chemical Society.]

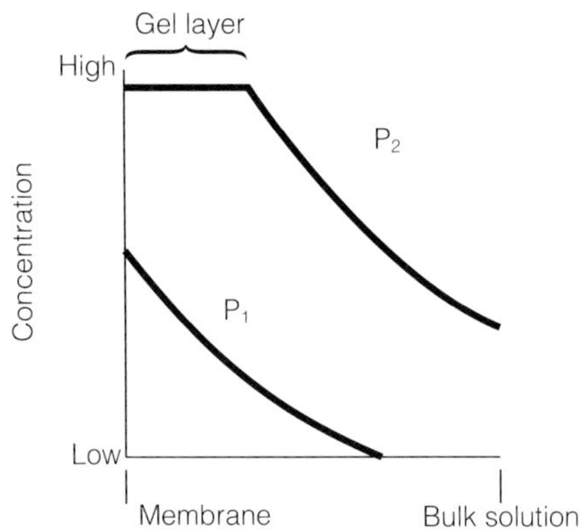

Figure 9.8. Schematic diagram showing the increased gel-layer thickness that forms when the transmembrane pressure P is increased and $P_2 > P_1$.

The flux of molecules across the membrane depends on the transport resistance through the gel layer. The presence of large proteins in the gel layer may hinder the transport of smaller molecules and prevent their passage through the membrane even though in the absence of the large molecule they would readily pass. The flux also increases with increasing pore size of the membrane, for nominal molecular–weight cutoffs greater than the weight of the transported molecule.

9.2.2 Fouling

The fouling of membranes may have a significant impact on performance in ultrafiltration. There are several reasons why the flux through the membrane may change during the process relative to the flux of pure water. These may be related to a change in the properties of the membrane, a change in the properties of the feed solution, concentration polarization, or fouling of the membrane. Membrane properties change as a result of chemical reaction, oxidation, embrittlement, and so on. Membranes may also be damaged by mechanical abrasion.

Concentration polarization may occur and worsen as the feed becomes more viscous or as its concentration increases. Concentration polarization should be reversible, but the flux may not always return to the original level. Fouling effects are usually irreversible. According to Cheryan (1986), "the general consensus appears to be that membrane fouling is due to the deposition and accumulation of submicron particles on the membrane surface and precipitation of smaller solutes within the pores of the membrane itself."

Fouling results in lower flux, shortened membrane life, and increased cost of operation. If the fouling can be predicted before the plant is designed, then it can be taken into account and sufficient surface area can be included to provide the necessary capacity after the fouling has occurred. The ultrafiltration of proteins is especially susceptible to fouling. Proteins are retained behind the membrane and build up in the gel–polarization layer. Under certain conditions proteins may precipitate and foul the membrane. Fouling can be avoided by maintaining conditions that minimize precipitation of the protein, such as the addition of chelating agents (for example, EDTA or citrate). The presence of salts would not be expected to influence fouling as the salts are small enough to pass through the membrane. Detergents, on the other hand, may form micelles that are larger than the pore size of the membrane and that may be concentrated along with proteins during the ultrafiltration process. These detergent micelles may contribute to gel polarization and fouling. Protein solutions, particularly cell culture media, may contain lipids, and these lipids may have an affinity, because of their hydrophobicity, for the membrane, which allows them to foul the membrane and reduce the flux.

As with concentration polarization, one might expect the shear rate to influence the rate of fouling. At higher shear rates the deposits may be scoured off the membrane and the flux improved. Back flushing through the membrane, when the membrane will withstand it, may also help to unclog and defoul the membrane.

A change in the properties of the feed solution may also affect the fouling. In particular, as the solution becomes more concentrated, the viscosity increases, the density increases, and the diffusivity tends to decrease—all leading to conditions that may promote precipitation and fouling. Several things can be done to minimize the effects of fouling. These include pretreating the feed, for example, with a chelating agent or some type of lipid–removing agent to reduce the tendency of these fouling layers to form or to prevent the reaction between the foulant and the membrane. It is also important to avoid particles that may plug the flow channels in the ultrafiltration device. Cheryan (1986) recommends, as a rule of thumb, a maximum particle size in the feed of one–tenth the smallest feed–channel dimension. Thus spiral–wound and hollow–fiber units need prefiltration down to 5 to 100 μm.

Feed treatments that increase the solubility of the material are advantageous because, as the material becomes more concentrated, the solubility will tend to decrease. Material that precipitates and fouls the membrane may become compressed and further reduce the flux as the pressure is increased. Proteins are less soluble near their isoelectric points. The flux tends to parallel the solubility, both having a minimum near the isoelectric point of the protein. Fouling aggravates the situation and leads to more rapidly reduced flux with increased pressure.

9.2.3 Materials

The first asymmetric membranes were developed by Loeb and Sourirajan about 1960. Asymmetric membranes typically have a structure with, on one side, a thin film having very small pore size, supported by a more open pore, or spongelike structure. This skin serves as the major resistance to mass transport through the membrane; and therefore the thinner the skin, the less is the mass–transport resistance.

Several types of polymer material are used for membrane filters. Cellulose acetate is made of beta–1–4–linked glucose units with one primary and two secondary hydroxyl groups and the beta glucosidic oxygen in the equatorial position. Cellulose acetate membranes vary in the degree of polymerization of the cellulose. The traditional source of cellulose is wood pulp or cotton linters. The advantages of cellulose acetate include high flux, relative ease of manufacture, and ready supply of raw material. The disadvantages include a limited temperature operating range (typically below 30°C, a disadvantage in terms of flux since higher temperatures lead to higher flux rates and a disadvantage in terms of sanitation since this temperature tends to breed microorganisms), a narrow pH range, typically pH 2 to 8 but preferably 3 to 6 (under more acidic conditions the polymer hydrolyzes, and under basic conditions deacetylation occurs, that is, at high and low pH the membrane degrades), and low resistance to sanitizing chemicals. Cellulose acetate is not resistant to chlorine and tends to deform or creep upon compaction. Because cellulose acetate can be degraded by microbial action and is difficult to sanitize with chlorine, its use is limited to nonsanitary conditions. It may still be applicable in one time use following gamma irradiation or gas sterilization. The polyamides, characterized by an amide bond (–CONH–) in their structure, have few benefits over cellulose acetate.

Polysulfone, made up of repeating diphenylene sulfone units, is widely used for ultrafiltration membranes and has several advantages. Temperatures up to 75°C are common, and some manufacturers claim temperatures higher than 125°C, which is an advantage for biotechnology applications where sterility is obtained by heat treatment at 121°C. Such membranes also have a wide pH tolerance; continuous exposure to pH 1 to 13 is not a problem. They have good resistance to chlorine with up to 200 ppm chlorine for short–term contact and 50 ppm for long–term storage. Polysulfone is easily fabricated in a wide variety of sizes and shapes, and a wide variety of pore sizes are available for ultrafiltration, with molecular–weight cutoffs from 1,000 to 500,000. Another type of filter is that of the ceramic membranes, which are gaining attention because of their rigid nature, high resistance to chemical treatment, and ease of production in a wide variety of configurations.

Two characteristics of pores of particular value when different membranes are copared are the pore–size distribution and the porosity. The pore–size distribution is the number of each size of pore in the membrane. The porosity refers to the total number of pores in a unit surface area of membrane. Ultrafiltration membranes are rated not by absolute pore size nor by the size of the particles that may go through them but rather by a nominal cutoff with reference to a comparable spherical protein having a certain molecular weight and its ease of passage through the membrane. These cannot be as rigorously set as in the microporous membranes because the protein they are trying to hold back is not so rigid as the particles used in rating microporous filters.

9.2.4 Configurations

The two configurations in which ultrafiltration membranes are in common use in industry are the hollow–fiber configuration and the plate–and–frame configuration. In the hollow fiber the feed solution is transported through the lumen of a small–diameter fiber (typically 0.2 to 1.5 mm in diameter, with a membrane thickness of about 0.2 mm). These are typically asymmetric membranes, with the skin on the interior, or lumen side, of the fiber. The process stream flows in the lumen down the length of the fiber, and the permeate travels through the membrane to the extracapillary space. The fibers are generally sealed in an epoxy resin at the ends, forming an arrangement that looks very much like a shell–and–tube heat exchanger. A manifold at each end provides inlet and outlet for the lumen, and outlet ports are also installed on the shell side. Hollow fibers typically operate with Reynolds numbers below 3,000 and are thus in the near–laminar– or laminar–flow region. High velocities in the lumen and the correspondingly high shear rates reduce the thickness of the boundary and gel–polarization layers. Of the available ultrafiltration designs hollow fibers have the highest area–to–volume ratio. Consideration must be given to the possibility that the pressure drop from one end of the fiber to the other (typically 35 to 140 kp, or 5 to 20 psig) may be greater than the pressure drop through the membrane of the inlet end and back through the membrane into the lumen at the outlet end. This may result in diversion of the flow out through the membrane, down the shell side, and back into the lumen through the membrane, rather than directly down the lumen. Hollow fibers have a low pressure rating, typically less than two atmospheres. Hollow fibers with a small diameter are susceptible to plugging if particles are present, so a clarified solution may be required.

Plate–and–frame ultrafiltration units come in many varieties, all with some features in common, namely, a manifold at the inlet side, flow across the membrane, and a manifold at the outlet end. A manifold is also present to collect the filtrate or permeate. Some type of spacer is required between the membranes to provide support and ensure that channels are open between the membrane surfaces. These support frames may have a tortuous, or open–channel, configuration. The tortuous channels provide higher shear rate and lower boundary–layer resistance but are prone to plugging if particles are present in the feed stream.

Another type of configuration is the spiral–wound cartridge, in which alternating membranes and spacers are wrapped around a central core. The feed comes in at one end, concentrate goes out the other, and the permeate or filtrate is caught between the spacers and transferred to an internal core. One disadvantage of the spiral–wound configuration is that the fluid permeate travels a long distance, resulting in a higher pressure drop and greater resistance to flux. Also, the long distance the fluid travels results in decreased flux owing to increased concentration in the feed, especially in the boundary layer. The spiral–wound modules typically have spacers giving tortuous paths. Reynolds numbers tend to be below 1,500, which is the laminar–flow region, but the spacers may create some turbulence. Spacers tend to create a high pressure drop, which may lead to a phenomenon known as "parasitic drag," causing the membrane to telescope. In most commercial modules some type of antitelescoping device is included to prevent the interior of the spiral from moving relative to the exterior of the spiral.

In the selection of an ultrafiltration design the effect of shear on the product may be important. If the protein is shear sensitive, then it may become denatured in a high–shear system. It is also important to consider this during the choice of a pump for operating the system, selecting one that has low shear, such as a positive–displacement rotary–lobe pump, instead of a centrifugal pump.

9.3 BUBBLE–POINT TESTING

The integrity of a membrane and the size of its pores can be determined by the bubble–point test. That a filter has integrity means that the liquid that passes through the membrane must pass through the pores of the membrane and cannot travel through the membrane through either tears or holes or circumvent the membrane by gaps between the membrane and housing. The bubble–point test is based on the action of liquids in capillaries. When a filter medium is wetted with the liquid and gas pressure is then applied to displace liquid out of the pores, the pressure that is required to blow the liquid from the pores is inversely proportional to the diameter of the pores. Thus a membrane passes the integrity test if the pressure reaches a certain minimum level before the liquid is blown out of the pores.

The test is based on Cantor's equation,

$$P = \frac{4\gamma(\cos\psi)}{d_\mathrm{p}} \tag{9.2}$$

where P is the bubble–point pressure, γ is the surface tension at the solvent–air interface, ψ is the liquid–solid contact angle, and d_p is the pore diameter. The test is conducted by thoroughly wetting the membrane with a liquid of known surface tension and contact angle and displacing the liquid by air. The pressure is increased until a steady stream of air bubbles is observed coming through the top of the membrane. This is the bubble–point pressure. A plot of the gas flow versus the differential pressure, as shown in Figure 9.9, clearly indicates the bubble point. Because the contact angle depends on both the liquid and the material of which the membrane is constructed, a given pore size will give a different bubble–point pressure

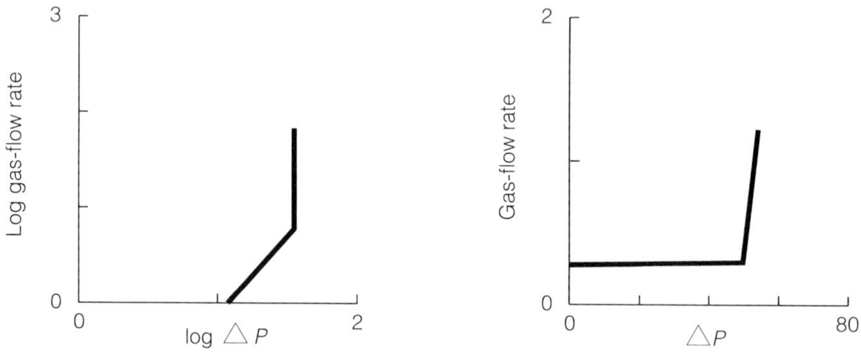

Figure 9.9. Plots of gas-flow rate as a function of differential pressure on both log and arithmetic scales. The discontinuity occurs at the bubble point. From Olson (1983); used with permission.

with different materials. According to Wolber and McAllister (1986) bubble–point determinations can be easily and accurately accomplished for surface areas up to 1,000 cm^2. Above this level the measurement becomes increasingly more subjective. Because of the high pressures required for bubble–point testing of ultrafiltration membranes, Cheryan (1986) suggests that the pore–size distribution of ultrafiltration membranes may be more accurately determined through the use of electron microscopy. As an alternative to increasing the pressure until bubbles pass the membrane, the pressure may be increased to a value below the bubble point, the system closed, and the change in pressure monitored with time. During the test gas will traverse the membrane by diffusion through the liquid in the pores. The rate of pressure loss due to diffusion will be much lower than loss due to breaks or gaps.

9.4 CLEANING AND SANITIZING

Filtration media must be clean before use, and media that are reused must be cleaned after use and in many cases sanitized as well. Microfiltration media are usually used only one time and therefore do not need postuse cleaning, while ultrafiltration media are designed to be reused many times.
Filter media may be sterilized by heat (live steam or autoclave) or chemical methods (ethylene oxide, sodium hydroxide, and the like). Pleated cartridges in their housings, for example, may be steam sterilized on line or off line. Membrane filters may not withstand the prolonged high temperatures necessary for heat sterilization but may be sanitized by treatment with hypochlorite (bleach) or hydroxide solutions. Small membrane filters are

often sterilized by ethylene oxide gas treatment. The choice of sanitizing method will depend on the membrane material and application.

Tangential–flow systems with reusable membranes should be cleaned immediately following use. It is important to avoid leaving protein residue on membranes in contact with air as removal of dried protein from membranes is very difficult. The membranes and attendant piping and tanks should be rinsed with water to remove the bulk of the remaining protein and to keep the membranes wet. In some cases buffer should be used instead of water to prevent precipitation of the protein on the membrane (see Chapter 8). Residual proteins may be removed by washing with sodium hydroxide and in severe cases with protein–digesting enzymes. For killing vegetative bacteria, Berglof et al. (1988) recommend 0.1 M NaOH for 60 minutes contact time.

9.5 APPLICATIONS

9.5.1 Concentration

Ultrafiltration is widely used to concentrate (dewater) protein solutions. Figure 9.10 is a diagram of the arrangement in cross–flow ultrafiltration for concentration or diafiltration of a protein solution. The solution to be concentrated is circulated from the reservoir through the membrane holder back to the reservoir. The filtrate that passes through the membrane is collected separately. The properties of the feed solution may change during the course of concentration. The viscosity typically increases, which may cause the pump to labor harder, raise the energy consumption, and increase the temperature of the material.

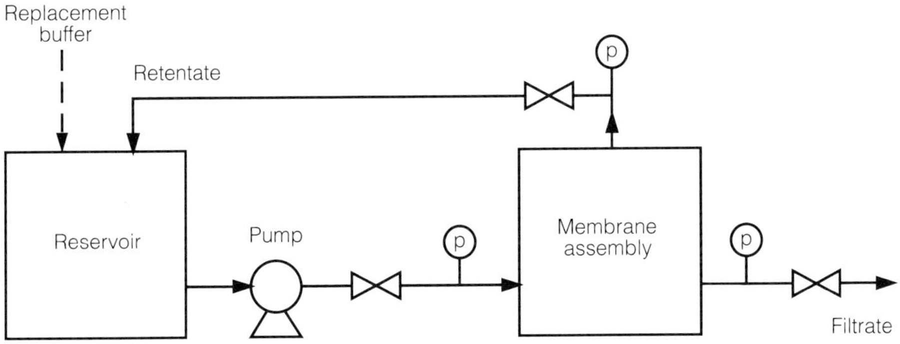

Figure 9.10. Diagram of process equipment for concentration or diafiltration (with replacement buffer).

9.5.2 Fractionation

The set up for fractionation is the same as for concentration. The goal of a fractionation process is to pass the molecules smaller than the product through the membrane or retain the molecules that are larger than the product behind the membrane. A general rule of thumb is that the two molecules to be separated need to differ in size by an order of magnitude and the ultrafiltration–membrane cutoff should be somewhere between the two particle sizes. For example, a 100,000–NMWCO membrane should effectively separate proteins of over 100,000 D from proteins under 10,000 D. Some portion of the proteins between 10,000 and 100,000 D would also pass the membrane.

9.5.3 Diafiltration

Ultrafiltration membranes are also used for diafiltration. This is a process with results identical with dialysis but operated under convective flow rather than straight diffusion. The purpose of this operation is to remove small–molecular–weight salt and buffer molecules associated with the protein and replace them with other low–molecular–weight species. This may be done in either continuous or discontinuous fashion. In the discontinuous mode the feed is concentrated, and the volume is increased by the addition of exchange buffer. This cycle is repeated (the solution is concentrated and additional exchange buffer is added) until the desired exchange is complete. In the continuous mode the exchanging buffer is brought in continuously, and the volume is held at a constant level during the operation.

In classical dialysis the driving force for diffusion of the salts across the membrane is the relative concentration of the salts in the two solutions separated by the membrane. In diafiltration the driving force for the transport of salt across the membrane is pressure. The salts in the starting solution are carried through the membrane by the streaming solvent. The salt leaving the solution is replaced by the exchanging buffer, which is added to the feed system. In dialysis the replacement buffer is on the opposite side of the membrane from the protein solution. In diafiltration the replacement buffer is added to the same side of the membrane as the protein solution. The actual mechanism of the two processes is somewhat different, but the result is the same.

One definite advantage of diafiltration over dialysis is the shorter process time of diafiltration: Diafiltration is rapid while dialysis is slow. Diafiltration is also more efficient in terms of the amount of exchanging buffer required to provide an equivalent exchange. Since diafiltration is a continuous process and uses a fixed volume, it is convenient to describe the efficiency of the operation in terms of the multiples of volumes of the material diafiltered and the percentage of salt removed. As a general rule of thumb, a sample volume turnover of three times will remove approximately 95 percent of the original salts, and a sample–volume turnover of five times will remove approximately 99 percent. For the case of continuous diafiltration this is described by the equation

$$\ln \frac{c_i}{c_f} = \frac{V_c - V'}{V_0} \tag{9.3}$$

where c_i is the initial microsolute concentration, c_f is the concentration of the final filtrate, V_c is the cumulative volume, V' is the holdup volume from membrane to filtrate outlet, and V_0 is the average sample volume.

The exchange of salt during a discontinuous diafiltration can also be easily calculated. In each step the solute level decreases according to the equation

$$\frac{c_f}{c_i} = \frac{V_f}{V_0'} \tag{9.4}$$

where c_f is the concentration after the concentration and dilution steps, c_i is the initial concentration, V_f is the sample volume after the concentration step, and V_0' is the sample volume after the concentration and dilution steps, which for the most part is equal to the initial sample volume. For example, if a protein solution containing 10 percent salt is concentrated to one–tenth the original volume and then diluted with buffer to bring it to the original volume, the salt concentration is reduced to 1 percent. Repeating this process would reduce it to 0.1 percent, and repeating it again would reduce it to 0.01 percent.

9.6 SCALE UP

Several factors must be considered in the selection of microfiltration media for large–scale protein purification, as summarized in Table 9.1. The volume of the process stream to be treated and the solids content of the stream will limit the choice of configuration and selection of depth or screen. If the product is in the liquid fraction, then protein binding must be considered. The sanitization method selected will depend on the process conditions as well as the filtration media selected.

Experiments should be conducted at the lab scale to determine the loading capacity (in units of liters of process fluid per unit area of membrane) of the

Table 9.1. Factors in filtration media selection.

Process-stream characteristics	
	a. Volume
	b. Solids or protein content
Media characteristics	
	a. Available configurations
	b. Solids-retention capacity or molecular-weight cutoff
	c. Structure (depth or screen)
	d. Sanitization method
	e. Protein binding
	f. Cost
	g. Dead end or cross flow for microfiltration; hollow fiber or plate and frame for ultrafiltration

selected filter and the flow rates and pressures of operation. Scale up consists of supplying sufficient capacity to accomplish the filtration in a reasonable time.

The length of time of a filtration operation bears on the level of pyrogens in the product. For example, if we assume a doubling time of 30 minutes per generation, a reduction in the filtration time by half, from 4 hours to 2 hours, will reduce the microorganism level by a factor of sixteen.

The two properties of ultrafiltration operations that most affect the cost are the flux and the retention. The higher the flux, the faster the process; the better the retention of solute, the more effective the separation and the smaller the equipment needed. The retention of solute is largely governed by the selection of the filtration medium. The flux is governed by the media, the process–stream characteristics (especially those properties that lessen concentration polarization and fouling), and the cross–flow velocity and transmembrane pressure.

Cross–flow ultrafiltration processes are scaled up by keeping the tangential velocity and transmembrane pressure constant. The optimal pressure gives the highest sustainable flux rate. At a given flow rate the flux increases with increasing transmembrane pressure to a maximum flux at the optimal pressure. Continued increase of pressure results in reduced flux. The maximum flux (optimal pressure) can be compared for several flow rates to determine the most favorable operating conditions for a given system.

The transmembrane pressure is the average of the inlet– and outlet–pressure differentials. That is,

$$P = \frac{P_i - P_o}{2} - P_f \tag{9.5}$$

where P_i is the inlet pressure, P_o is the outlet pressure, and P_f is the filtrate pressure.

9.7 OTHER MEMBRANE PROCESSES

In addition to microfiltration and ultrafiltration other membrane processes that find limited application in large–scale protein purification include electrodialysis, reverse osmosis, and perm–selective membranes. Table 9.2 compares these membrane processes in terms of application, driving force, and membrane–pore size.

Electrodialysis removes small ionic species from process streams. The apparatus consists of alternating anion and cation exchange membranes (isotropic membranes with charged groups attached) separated by spacers containing flow channels. An electric field is produced by a pair of electrodes, positive on one side and negative on the other. Positive ions pass through the cation exchange membranes, negative ions pass through the anion exchange membranes; the ion–enriched stream is separated by a manifold from the ion–depleted stream. An example application is the removal of salt from protein streams in cheese making.

Reverse osmosis removes water from process streams by membranes having pores too small to allow passage of even simple salt ions. If two solutions of different solute concentration are separated by a membrane permeable to the solvent, solvent molecules will pass from the dilute compartment to the

Table 9.2. Comparison of membrane processes.

Membrane process	Moieties separated	Driving force	Pore size
Microfiltration	Large particles	Pressure (< 70 kp)	> 0.1 μm
Ultrafiltration	Large soluble molecules	Pressure (< 350 kp)	> 0.01 μm
Reverse osmosis	Small soluble molecules	Pressure (< 7000 kp)	
Electrodialysis	Charged species	Electric field	
Perm-selection	Selected species	Concentration difference	

concentrated compartment until the activity within the two compartments is equal. This driving force toward dilution can be counteracted by the application of pressure to the concentrated compartment, driving the solvent (water) through the membrane to the dilute side. Reverse osmosis is used mainly for the production of high–purity water.

Perm–selective membranes allow the transport of specific molecules while excluding all others. Solute molecules are dissolved in the membrane and transported by diffusion to the far side. An example would be the use of a chelating membrane for removal of metal ions from a process stream.

REFERENCES

Berglof, J. H., N. P. Adner, and S. Y. Doversten, "Inactivation of microbial contamination in chromatographic separation media using sodium hydroxide," paper presented at the XX Congress of the International Society of Blood Transfusion, London, July 10–15 (1988).

Bruin, S., A. Kikkert, J. A. G. Weldring, and J. Hiddink, "Overview of concentration polarization in ultrafiltration," *Desalination* 35:223–242 (1980).

Cheryan, M., Ultrafiltration Handbook, Technomic Publishing, Lancaster, PA (1986).

Olson, W. P., "Process Microfiltration of Plasma Proteins," *Process Biochemistry* 18:29–33 (1983).

Porter, M. C., "Concentration polarization with membrane ultrafiltration," *Industrial and Engineering Chemistry Product Research and Development* 11(3):234–248 (1972).

Pusch, W., "Membranes from cellulose and its derivatives," in Wood and Cellulosics, A. Kennedy (ed.), Ellis Horwood, Chichester, England (1987).

Pusch, W., and A. Walch, "Synthetic membranes: preparation, structure, and application," *Angewandte Chemie International Edition* (English) 21:660–685 (1982).

Trudel, M., P. Trepanier, and P. Payment, "Concentration and analysis of labile viruses by hollow fibre ultrafiltration and ultracentrifugation", *Process Biochemistry* 18(1):2–4, 9 (1983).

Wolber, P. M., and M. A. McAllister, "Bubble–point determinations: a comparison of manual vs automated methods," Fluid Filtration: Liquid, vol. 2, ASTM STP 975, P. R. Johnston and H. G. Schroeder (eds.), American Society for Testing Materials, Philadelphia (1986).

FURTHER READING

Johnston, P. R., "Another view of the integrity test, or bubble–point test, applied to fluid filter media," in Fluid Filtration: Liquid, vol. 2, ASTM STP 975, P. R. Johnston and H. G. Schroeder (eds.), American Society for Testing Materials, Philadelphia (1986).

Lonsdale, H. K., "The growth of membrane technology," *Journal of Membrane Science* 10:81–181 (1982).

APPENDIX 9 Vendors of filters, membranes, and associated equipment.

Amicon 24 Cherry Hill Dr. Danvers, MA 01923 800–343–1397	Ultrafiltration
Filtron Pharmacia LKB Biotechnology 800 Centennial Ave. Piscataway, NJ 08854 800–922–0318	Ultrafiltration
Gelman Sciences Inc. 600 S. Wagner Rd. Ann Arbor, MI 48106 800–521–1520	Microfiltration and ultrafiltration
Millipore Corporation 80 Ashby Rd. reverse osmosis Bedford, MA 01730 617–275–9200	Microfiltration, ultrafiltration
Pall Process Filtration Co. Northern Blvd membranes East Hills, NY 11548 516–484–5400	Microfiltration and activated 2200
Sartorious GmbH P.O. Box 3243 Weender Landstrasse 94–108 D–3400 Goettingen Federal Republic of Germany 551–308–1	Microfiltration
Sepracor 33 Locke Dr. Marlborough, MA 01752 508–481–6700	Activated membranes

CHAPTER 10
GENERAL CHROMATOGRAPHY

Of all the purification methods available for separation of proteins, perhaps the most useful is chromatography, with its wide range of bases for separation and of conditions for operation. Chromatography is a separation method in which the different components in a sample migrate through a column at different rates.

There are basically two mechanisms for chromatography: adsorption, in which the sample molecules are adsorbed onto the chromatographic medium through binding (either noncovalent in the case of ion exchange and hydrophobic interaction or covalent in the case of thiophilic affinity chromatography), and nonadsorption (such as gel filtration chromatography). In adsorption chromatography the rate of migration of the sample is controlled by the thermodynamic properties of the system (adsorption characteristics), and the shape of the elution profile is determined by kinetic factors (diffusion characteristics). In gel filtration chromatography, for a given sample and resin, the rate of migration is determined by kinetic factors.

This chapter deals with those features which are common to all types of chromatography, with an emphasis on adsorption chromatography; differences among the various methods are detailed in the next several chapters.

Many different interactions can be exploited as bases for chromatography, including ionic interaction, hydrophobic interaction, and size as well as specific interactions, as in affinity chromatography. Types of adsorption chromatography are classified by the interaction that occurs between the sample and the matrix. Ion exchange chromatography is based on electrostatic interactions between molecules in the sample and the ion exchange matrix. Hydrophobic interaction and reversed phase chromatography are based on interactions between hydrophobic portions of the sample and a hydrophobic solid phase. Affinity chromatography is based on specific interactions between the sample and a complementary structure bonded to the solid phase. Gel filtration chromatography is based on the different rates of diffusion of molecules of different sizes in a porous matrix.

In this book chromatography is described in the following terms. Chromatography consists of interactions between molecules in the sample (fluid phase) and the matrix or gel medium (solid phase). The matrix is often packed in a column (fixed bed). The sample is eluted, or recovered, from the matrix by the action of an elution buffer or solvent. The elapsed time from the initial application of the elution buffer until the exit of the sample peak from the column is the retention time; the volume of elution buffer applied in this time is the elution volume. The eluted fractions are also known as the eluate. The record of the conductivity or absorbance of the eluate over time is known as the chromatogram.

10.1 GENERAL THEORY

Adsorption chromatography is based on the distribution of a sample molecule between two phases with which it is in contact. Consider, for example, a species initially at equilibrium in the fluid phase. When brought into contact with a solid phase, this species will be distributed between the solid and liquid phases according to its affinity for the matrix; molecules with greater affinity for the solid phase will have higher concentrations in the solid phase, and lower concentration in the fluid phase. The term "affinity" here refers to the attraction between the sample molecule and the solid phase. The affinity of multiple species can be compared *relative* to each other. The distribution of the component between the two phases depends on several factors that include the physical interactions, chemical interactions, and combined forces between the component and each phase. Because different components have different distribution characteristics (as a result of different physical characteristics), it is possible to separate the components from each other.

Chromatography is a batch operation consisting of two steps: sample application and sample elution. In some cases (ion exchange and some affinity methods) an additional regeneration step is required. In the first step the sample is adsorbed to the solid phase; in the second step the components of the sample are resolved into distinct bands in the column and eluted as separate fractions. If necessary, the column is then regenerated to return it to a form to which the next sample can be applied. Figure 10.1 illustrates these different steps. The four columns represent snapshots of a single chromatography column taken at four different times, to capture the position of the sample species as they move through the column. In the leftmost column, at time 0, the sample, containing multiple species (represented by squares, triangles, and circles) is applied to the top of the column. As time progresses to 1 and 2, the components of the sample become ordered so that all of one species is within the same cross–sectional plane of the column. Eventually, at time 3, the components are eluted from the column and each

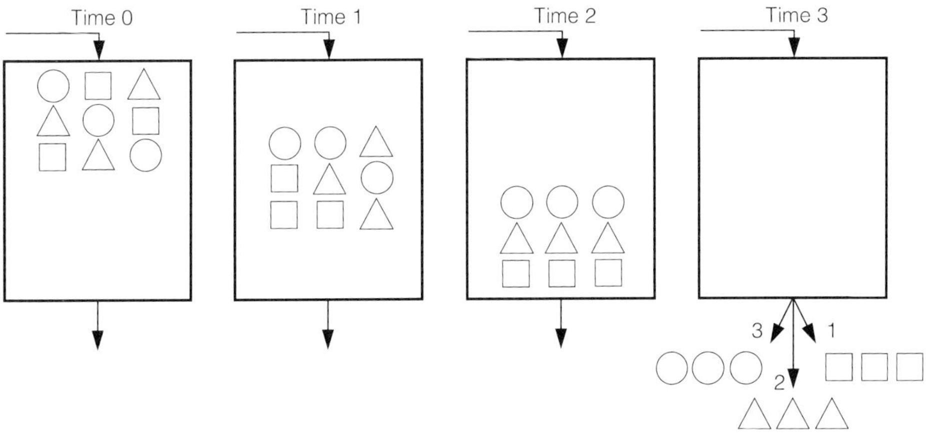

Figure 10.1. Schematic diagram of chromatography, showing the relative movement of species through a column with time.

species collected separately, that is, all circles together, all squares together, and so on.

Chromatography systems, as other systems in nature, are often represented by one of two types of mathematical model. The first type is the design model, so called because it predicts behavior and can be used to design a real system (such as the elution time and the shape of the elution curve). The second type is not used for design but rather to illustrate what goes on at the fundamental level (such as the electrostatic interaction between protein and solid phase). Most models in chromatography are design models, and these can be further divided into two classes: those which posit a specific mechanism (mass transfer, local equilibrium, and so on) and include column parameters (such as particle size or sample volume) and those which are independent of the actual mechanism (for example, statistical models). A mechanistic model that fits empirical data does not necessarily represent the true mechanism.

In the case of a mechanistic model for adsorption chromatography two parts are needed to describe the system fully: one for the equilibrium distribution of sample between the phases and one for rate (mass–transport) phenomena. In the case of gel filtration the rate model alone is sufficient (since there is no equilibrium for adsorption to the surface). The first of these two parts, equilibrium partitioning, results from thermodynamic properties of the sample and matrix that control the elution time and relative position of components. The second part, the mass–transfer characteristics of the system, controls the shape of the elution curve.

The rate of mass transfer for proteins has significant impact on column behavior because of the size of the molecules, which leads to low diffusivity. The diffusion coefficient for polymers at room temperature is of the order of 10^{-7} to 10^{-10} cm^2/s, or one–hundredth that of small molecules. For example, the diffusion coefficient for myosin (molecular weight 440,000 D) is approximately 10^{-7} while that of glycine (molecular weight 75 D) is approximately 10^{-5}. The variation of diffusion coefficient with molecular weight is given by the Stokes–Einstein equation as

$$D \alpha \text{ (molecular weight)}^{-n} \tag{10.1}$$

where n = 1/3 to 1/2.

Models used to describe preparative chromatography can be grouped into three classifications: equilibrium models (which ignore mass–transfer resistances), rate models (which include mass–transfer effects), and statistical models. Equilibrium models will be described first, followed by models that include rate phenomena, followed by statistical models.

10.1.1 Equilibrium Models

The equilibrium distribution of a molecule between two phases can be represented by a simple equation. Since this distribution varies with temperature, the distribution at a given temperature is called an "isotherm." The simplest form is the linear isotherm,

$$K = \frac{x}{y} \tag{10.2}$$

where K is the thermodynamic equilibrium constant, x is the concentration in the mobile phase, and y is the concentration in the solid phase. Other distribution models that are commonly used are the constant separation factor, mass action, Langmuir, and Freundlich isotherms, listed in Table 10.1. An example of the Langmuir–type isotherm is shown in Figure 10.2.

Note that the lower–left part of the Langmuir isotherm, which represents low concentrations, is linear but that at higher concentrations the isotherm becomes nonlinear. Analytical chromatography is invariably run in the linear region; preparative chromatography is often run in the nonlinear region, where the retention time depends on the quantity of sample applied to the column. Linear isotherms apply only if the interactions between molecules in the sample are negligible.

Table 10.1. Common isotherm models (from Vermeulen et al., 1973).

Model	Solid-phase concentration	Fluid-phase concentration
Linear	x	y
Constant separation factor (R)	$\frac{x}{R + x - Rx}$	$\frac{Ry}{1 + Ry - y}$
Langmuir	$\frac{x(1 + Kc_r)}{1 + Kc_r x}$	$\frac{y}{1 + Kc_r(1 - y)}$
Freundlich	x^{δ}	$y^{1/\delta}$
Mass action	$K\left(\frac{C_0}{Q_0}\right)^{k-1} = \left(\frac{y_A}{x_A}\right)^{k}\left(\frac{x_B}{y_B}\right)$	

The selectivity factor α is the ratio of the equilibrium constants of two components. The selectivity factor can be altered by changing the nature of the mobile or solid phases. A small increase in α may greatly simplify the separation, while a small decrease in α may greatly increase the difficulty of separation.

The composition at any point within the column can be calculated by solving differential equations representing material balances for the components, mobile phase, and solid phase. The differential material balance, as given by Vermeulen et al. (1984), is of the form

$$\frac{DA\varepsilon}{Q}\frac{\partial^2 c}{\partial v^2} = \frac{\partial c}{\partial v} + \rho_s \frac{\partial q}{\partial V} + \varepsilon \frac{\partial c}{\partial V} \tag{10.3}$$

where c and q are the fluid phase and resin phase concentrations, v is the volume of sorbent from the inlet to the point of interest (or the distance variable), V is the volume of fluid fed to the column (or the time variable), ρ_b is the sorbent density, D is the diffusivity, A is the cross–sectional area, Q is the volumetric flow rate, and ε is the void fraction. The term on the left represents dispersion effects, the first term on the right represents convection, and the second and third terms represent accumulation in the resin and fluid phases. The rate at which equilibrium is reached depends in part on the rate of mass transfer between the phases. The rate expression is of the form

$$\frac{\partial q}{\partial V} = \kappa f(c, q) \tag{10.4}$$

where κ represents the mass transfer coefficient and $f(c,q)$ the driving force, that is, the equilibrium distribution isotherm. If the rate of mass transfer is assumed to be infinite, the equilibrium isotherm can be substituted directly into the differential material balance. This is the basis for equilibrium models of chromatography.

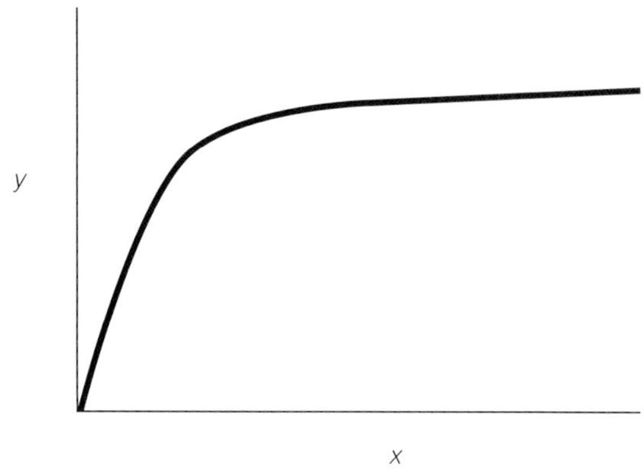

Figure 10.2. Representative isotherm (Langmuir type), illustrating distribution of component between fluid phase (x) and solid phase (y).

The assumption of a linear isotherm allows equilibrium models to be solved analytically. More complicated isotherms can be solved numerically for small numbers of components. These models suffer from oversimplification of the process and fail to predict the effect of changes in operating conditions.

If the rate of mass transfer is considered to be finite, then the material balance and the rate equation must be solved simultaneously. This is the basis of rate models.

10.1.2 Rate Models

The realities of finite mass–transfer rates mean that for practical purposes equilibrium between the mobile and solid phase is never quite reached. The effects of finite mass–transfer rates are exhibited in band broadening or band spreading: the deviation of the elution profile from the equilibrium ideal. In addition, band spreading results from longitudinal (axial) dispersion and such packing or flow irregularities as channeling.

The rate models work to solve the material–balance equations along with boundary and initial conditions but without any assumptions of equilibrium between phases. Much of the work in rate theories has been directed toward fixed–bed adsorption and ion exchange under conditions of "frontal analysis" or "breakthrough," where the sample is continuously applied to the column until it begins to appear in the effluent.

Variations between rate models are mainly due to the use of different equations to describe the mass–transfer rate. Resistances exist in several places between the mobile and solid phases, and have been described by Yang and Tsao (1982) as in the fluid exterior to the particles (film diffusion rate controlling), within the fluid of the particle pores (pore diffusion rate controlling), and on the surface of the pore (surface adsorption rate controlling), and as due to finite reaction times (reaction rate controlling). The rate may be controlled by any or several of these.

The weakness of rate models is the complexity of the mathematical equations, which are often so intractable as to be insolvable. In particular, the variation of the concentration through the particle is difficult to handle and is often simplified by the assumption that the interior concentration is uniform. However, the rate models may provide a qualitative idea of how changes in parameters, such as particle size or pore size, affect column performance.

10.1.3 Statistical Models

In addition to the equilibrium and rate models discussed above, statistical formulas have been derived that model or quantify chromatographic performance and allow comparison of experimental results. Foremost among these is the statistical–moments theory of Giddings and Eyring (1955), in which statistical moments are used to characterize the elution peak. The moment of a peak is a mathematical equation that describes a characteristic of the distribution function.

The first moment relates to the retention time, the second moment to the peak width, and the third and fourth moments to the symmetry and flatness or peakedness (kurtosis) of the peak. These moments are defined by Gareil and Rosset (1982) as follows: The zero–order moment is the quantity injected V_i,

$$M_o = C(V)\,dV = V_i \tag{10.5}$$

The normalized first–order moment is equal to the elution volume

$$M_1 = \left(\frac{1}{M_o}\right) C(V)\,V\,dV = V_e \tag{10.6}$$

The normalized higher–order moments are

$$M_i = \left(\frac{1}{M_o}\right) C(V)\,(V - M_i)^i\,dV \tag{10.7}$$

The second–order moment measures the variance of the profile ($M_2 = \sigma^2$)

$$\mathrm{NTP} = \frac{M_1^2}{M_2} = \frac{V_e^2}{\sigma^2} \tag{10.8}$$

where *NTP* is the number of theoretical plates (see below) and σ is the standard deviation or mean quarter width of a Gaussian peak. As defined here, Equations 10.5 and 10.6 are absolute moments (that is, they are taken with respect to t=0), and Equation 10.7 is a central moment (that is, taken with respect to M_1). Thus M_1 may be termed the first absolute moment and M_2 the second central moment.

The stochastic model, also known as the random–walk model or molecular dynamic model, is also based on the work of Giddings and Eyring (1955). In this model a probability function of the Poisson type is derived to describe the distribution of residence times of a sample molecule. Interactions between molecules are not included, limiting this model to linear isotherm systems, and the rate–controlling step is assumed to be the kinetics of the adsorption and desorption reactions with mass–transfer limitations neglected.

Because the chromatography of proteins is too complicated for representation by simple equations, simplifying assumptions are incorporated to reduce the level of complexity. These assumptions are necessary either to make the equations tractable or to cover for insufficient information about the sample or matrix. The assumptions frequently simplify the problem to the point where the model becomes a poor predictor of actual performance.

Simple cases have been solved by exact solution of the differential equations. Slightly more complex cases have been solved numerically by computer. Complex cases involving mixtures of proteins have not been solved. Protein mixtures present many problems to the modeler: Very little data on the thermodynamics of protein adsorption are available; proteins interact with each other in solution, and therefore the measurement of adsorption isotherms for single components will not reflect the behavior of a mixture; the starting material for a chromatographic separation contains tens of different proteins, each having different adsorption characteristics. Without detailed thermodynamic data it is impossible to get meaningful results from kinetic or equilibrium models. No simple techniques exist for obtaining the isotherms of complex protein mixtures.

10.1.4 Measures of Resolution

The failure of models to satisfy the needs of chromatographers, either because simple models are too inaccurate or because accurate ones are too difficult to solve, has led to the use of other measures of column performance. These alternatives do not necessarily predict elution profiles from fundamental properties. Rather, they serve as a means of interpreting column data and provide a basis for comparing results from different operating conditions.

The simplest measures merely quantify the resolution in terms of retention times and peak widths. For example, we may define resolution by the following equation:

$$R = \frac{2(t_2 - t_1)}{w_2 - w_1} \tag{10.9}$$

that is, the difference in times of retention of two peaks divided by the average of the width of the elution curves at the baselines of the two peaks. The retention time is a function of flow rate and column geometry.

Another useful measure is the retention volume or elution volume

$$V_e = V_v + K(V_t - V_v) \tag{10.10}$$

where V_v is the column void volume, V_t is the total column volume, and K is the distribution constant or the ratio of concentrations of solute in the solid and mobile phases. The void volume is that portion of the column volume that is not occupied by the chromatographic medium; for closepacked spheres this is 0.37, and for compressible or irregularly shaped media the void fraction is somewhat less. The term $V_t - V_v$ represents the volume of the solid phase that is available for chromatography. The retention volume divided by the volumetric flow rate gives the retention time.

A common and useful measure of performance is the number of theoretical plates. The phrase is borrowed from the petroleum–distillation industry, where it originally applied to columns that contained discrete plates, but it has since been applied to packed columns. The number of theoretical plates is the number of plates having 100–percent efficiency that are required to give a desired separation. (The actual number of plates is always greater than the theoretical number due to real–life inefficiencies.) For packed columns the height of packing that is equivalent to one theoretical plate, or "height equivalent to a theoretical plate" (*HETP*) provides a comparative value. Though the *HETP* is borrowed from the equilibrium–plate model, the inadequacies of the model do not negate the value of the *HETP* as a measure of column performance.

The *HETP* is an empirical determination and therefore strictly applies to only one particular column size, packing material, and flow rate. In spite of these limitations, it provides a useful way to compare the performance of a column over time. The length of the column divided by the number of theoretical plates *NTP* equals the *HETP*

$$NTP = \frac{\mathrm{L}}{HETP} \tag{10.11}$$

The value of *NTP* is calculated from the elution volume or retention time and the peak width or standard deviation:

$$NTP = 16\left(\frac{V_{\mathrm{e}}}{w}\right)^2 = \frac{\mathrm{t}^2}{\sigma^2} \tag{10.12}$$

where V_e is the elution volume and w is the width of the peak at the baseline. The units of V_e and w must be the same (liters). The peak width at the baseline can be measured by extending tangents from the sides of the peak, or the width at the midheight of the peak can be used if the value of the constant is changed from 16 to 5.54. Because the absolute value of the *HETP* is dependent on several column parameters, a reduced value h equal to the *HETP* divided by the particle diameter, is sometimes used.

The resolution one can expect from a column is influenced by the number of theoretical plates; the larger *NTP* is, the better the column is packed and the better the resolution. Typical low–pressure (atmospheric–pressure) columns may have on the order of a few thousand plates per meter, while high–performance (pressures up to 3,500 kp, 500 psi, or more) columns are now available that offer upward of 50,000 theoretical plates per meter of column length.

The *HETP* is often measured in preparative columns with a small (relative to production size) sample, 1 percent of column volume or less. In gel filtration columns a salt sample is sometimes run instead of protein. A low number of theoretical plates and an asymmetric peak indicate poor column packing. An asymmetric peak that shows a long tail on the upstream side may be the result of channels or cracks in the longitudinal direction through the bed that allow the sample to migrate unevenly. Tailing on the downstream side of the peak may be due to uneven packing with a section of the bed compressed more densely than the rest.

10.2 RESOLUTION

Perhaps the most critical issue in chromatography is resolution, or the quality of the separation measured in terms of the purity of the component fractions. The higher the resolution, the less overlap between species and the greater the purity of each species.

The many factors that affect resolution, as diagrammed in Figure 10.3, may be divided into classifications similar to those used to describe the models. On the equilibrium side we have reaction rate and reaction efficiency; on the mass–transfer side we have diffusion between the phases (through the film and within the pores), axial diffusion, and energy losses in the hardware. We shall discuss the factors that affect resolution in terms of their physical location: in the fluid phase, in the sample, or in the solid phase. The factors that stem from hardware are included in a later section on equipment.

10.2.1 Fluid Phase

The factors in the fluid phase that most affect resolution are related to temperature and flow rate. The gradient profile, which plays a lesser role, is described in a later section, Practical Operations.

As the temperature at which the operation is conducted is decreased, several changes occur. First, the viscosity of the fluid increases. This hinders mass transfer between the liquid and solid phases. Second, the diffusivity of proteins decreases, also hindering transport of the molecules between the liquid and solid phases. The increase in viscosity (for aqueous buffers, from 0.9 cP at 25°C to 1.5 cP at 4°C) increases the pressure drop. Buffer pH may vary with temperature, particularly for solutions buffered with tris, where a decrease in the temperature increases the pH about 0.03 pH units per degree Celsius in the range from 25 to 5°C. Because of these effects, it is frequently preferable to operate a quick process at room temperature rather than a slow process at a reduced temperature. Vanecek and Regnier (1980) reported a 40–percent decrease in resolution as the temperature was lowered from 25 to 4°C for the purification of albumin.

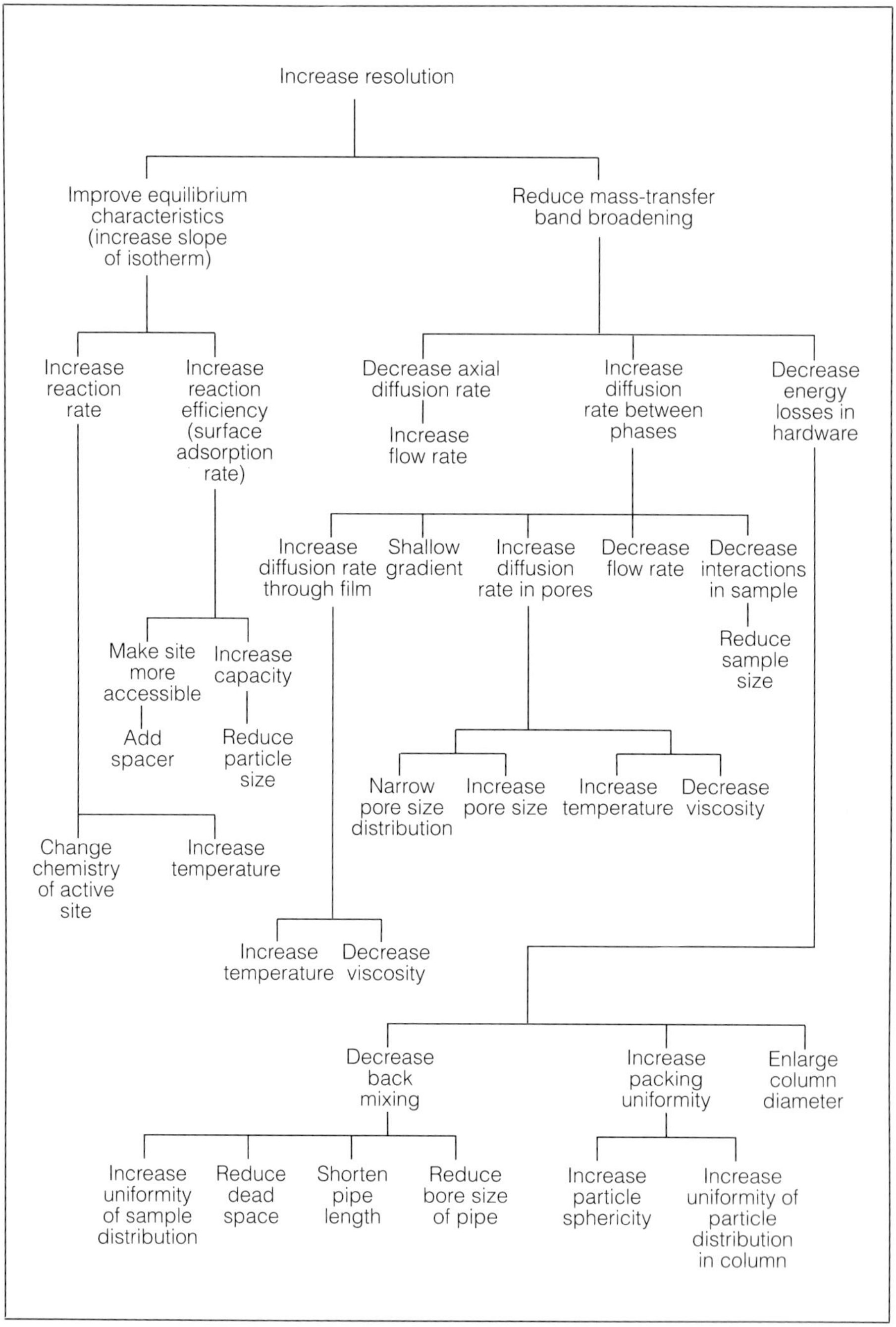

Figure 10.3. Morphological diagram of factors that increase resolution in chromatography.

The flow rate of the elution buffer affects the quality of the separation and the extent of resolution. Maximum resolution occurs when the solution moving through the column is in constant local equilibrium with the solid phase, in the absence of other detracting mechanisms, such as longitudinal diffusion. Because mass–transfer resistances exist between the fluid and solid phase, the material is not transported instantaneously between the two phases and equilibrium is not instantaneously attained. At high flow rates the departure from equilibrium is greater than at low flow rates as a result of these mass–transfer resistances.

Vanecek and Regnier (1980) have pointed out that, since the resolution is not directly proportional to the mobile–fluid flow rate, a decreasing flow rate is not always advantageous for increasing resolution; and since the separation time increases with a decrease in flow rate, some point is reached where it is no longer advantageous to reduce the flow rate. The approach to local equilibrium increases with decreasing flow rate while the longitudinal or axial diffusion decreases with increasing flow rate. The optimal flow rate is described in the plot of Van Deemter et al. (1956), shown in Figure 10.4, as the flow rate at which the plate height is a minimum.

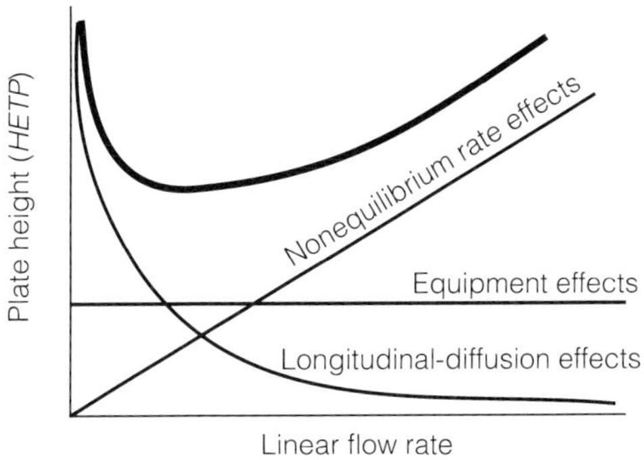

Figure 10.4. Cumulative effects of dispersive factors on plate height as a function of flow rate. (Reproduced by permission of Springer Verlag.)

In actual practice the effects of local equilibrium play a greater role than do those of longitudinal diffusion. Figure 10.5, from Regnier and Gooding (1980), shows the increase in plate height with flow rate for samples with different molecular weights. The absence of a minimum indicates that diffusion between mobile and solid phases is much more important than diffusion in the axial direction.

The optimal time with respect to resolution may not be the optimal time with respect to cost. Figure 10.5 suggests that the slower the flow rate, the higher the resolution. But slowing the flow rate increases the time of the operation and consequently increases the cost of the process. As described by Giddings (1965), the shortest possible time of separation for any point on the Van Deemter curve is proportional to the slope of the chord between that point and the origin. The lesser the slope of the Van Deemter curve,

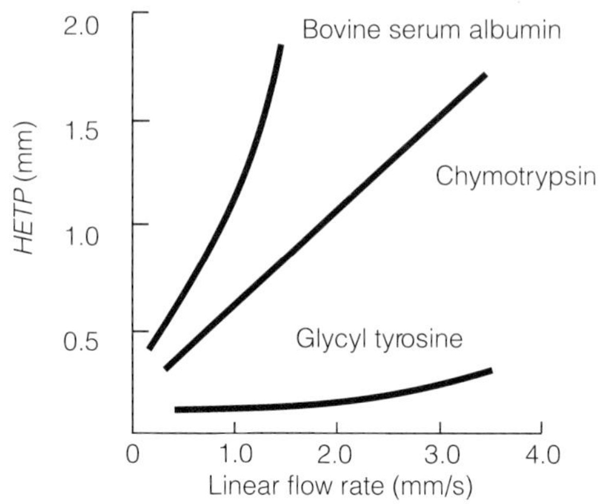

Figure 10.5. Plate height as function of flow rate for three different species. (Reproduced by permission of Academic Press.)

the greater the advantage gained with respect to a shortened process time from increasing the flow rate. From Figure 10.5 one can see that in the case of bovine serum albumin, increasing the flow rate from 0.5 mm/s to 1 mm/s doubles the plate height while the slope of the chord from the origin to the curve increases hardly at all and that the time required for the separation would be little changed at great cost to resolution. A similar change in the flow rate for glycyl tyrosine gives little increase in plate height, with a substantial change in the slope of the line between the origin and the curve and greatly reduced separation time with little loss of resolution.

10.2.2 Sample

The sample volume and the sample concentration both affect the resolution in preparative chromatography. At low concentrations the isotherm remains linear, and as the volume is increased, the peak becomes wider. As the protein concentration is increased, the peak height increases until the distribution isotherm becomes nonlinear owing to interactions between molecules in the sample, and the peak shape tends to skew. Knox and

Pyper (1986) suggest that up to some critical ratio of sample volume to column volume the profile of the peak is independent of the sample volume but that beyond that ratio resolution tends to decrease.

Vanecek and Regnier (1980) found that high sample concentration caused a decrease in resolution. The concentration of protein samples in preparative chromatography is typically 0.5 to 4 mg of protein per milliliter of column, although the concentration may be as high as 10 mg of protein per milliliter of column or higher.

If the viscosity of the sample is significantly different from the viscosity of the elution buffer, it will be difficult to maintain an even distribution of sample across the column.

10.2.3 Solid Phase

The matrix parameters that affect column resolution include particle shape, particle size, particle–size distribution, pore diameter, pore size distribution, capacity, surface area, pore volume, and surface activity (specific and nonspecific binding sites).

Spherical particles give more uniform packing and usually better performance. Particles of irregular shape are less expensive to produce; this is particularly true for ceramics, including silica. Decreasing the particle size increases the resolution by providing a greater number of accessible sites. Media of narrow particle–size distribution are usually produced from broad distribution media through classification by sieving or air buoyancy separation. A medium with a wider particle–size distribution costs less than one with a narrower distribution. The substitution of media with broader particle–size distribution will usually reduce the resolution of the column.

As the particle diameter decreases, efficiency, or the total number of plates (the column height divided by the *HETP*), increases. It is this change in particle size which gives high–performance liquid chromatography its advantage over atmospheric–pressure columns, namely, the much greater number of theoretical plates that can be packed into a column of reasonable length. Of course, the smaller the particle size, the higher the pressure drop per unit length. The pressure drop increases with the inverse of the square of the particle. Thus a reduction in particle size from 120 to 40 μm requires a ninefold pressure increase to obtain the same flow rate. This increased pressure drop may have a significant impact on the cost of operating the process. The cost of media also increases with decreasing particle size.

The pores must be of sufficient diameter to permit access by proteins—typically greater than 20 nm—and, according to Kopaciewicz et al. (1987), greater than 50 nm for proteins of molecular weight above 100,000 D. A narrow pore–size distribution is generally preferred, as a wide distribution may include pores too narrow for effective access by proteins. In gel filtration chromatography a wider pore size distribution fractionates a wider range of molecular weights, while a narrower distribution increases the resolution between molecules of similar size.

The capacity of the medium in chromatography (except gel filtration) is determined by the number of available adsorption sites, the majority of which lie on the surface of the pores. Only those pores that are large enough to permit access by the protein contribute to capacity. The greater the number of pores, the greater the available surface area for adsorption. An increase in surface area increases the capacity only to the extent that the area within pores that are large enough to permit the entry of proteins is increased.

The pore volume plays a secondary role in adsorption chromatography but is very important in gel filtration chromatography. A change in the pore volume will alter the performance of a partition step only if it leads to a change in the accessibility of the surface to the proteins. In gel filtration chromatography, however, a change in pore volume may alter the molecular–weight range of proteins that can be fractionated. A greatly increased pore volume may give a weaker matrix structure.

The surface of the matrix in partition chromatography contains active sites that selectively adsorb protein. The number of these sites within pores accessible to protein determines the capacity of the adsorbent. The surface may also contain minor sites that lead to nonspecific adsorption, for example, hydrophobic sites on an ion exchange matrix. Variation in the concentration of specific and nonspecific sites is responsible for much of the performance differences between manufacturers and even between lots from the same manufacturer.

10.3 PRACTICAL OPERATION

The three modes of elution by which a sample is recovered from a chromatography column are isocratic elution, in which the equilibration buffer is used for elution; gradient elution, or a change in the relative affinity of the sample and the matrix, the most common mechanism in analytical chromatography; and displacement of the sample by a substance of greater affinity for the matrix, a mechanism that occurs in preparative chromatography more often than is probably recognized. These three modes are contrasted in Table 10.2.

Table 10.2. Characteristics of the three elution modes.

	Isocratic	Gradient	Displacement
Isotherm	Linear	Tends to be linear	Non-linear
Sample size	Small	Small to moderate	Large
Separation basis for multiple components	Slopes of isotherms of components differ	Varying affinity between sample and solid phase	Displacement of lesser affinities to front
Travel rate of bands	Bands travel at different rates	Only band with affinity equal to eluting agent travels	Bands travel at same rate (i.e., they are coherent)
Affinity of elution agent for matrix	Equal to or less than sample	Varies with time	Greater than sample
Boundaries	Non-sharpening	Non-sharpening	Self sharpening
Component interaction	None	Usually none	May be severe

In isocratic elution the relative affinity between the molecules being separated and the chromatography matrix must be low so that the molecules may be easily displaced from the matrix by the same buffer used to equilibrate the column. When a sample is added to such a chromatographic system and the buffer is applied, the components migrate through the column at rates depending on their affinity for the matrix. In isocratic elution there is no change in the affinity of the components for the resin during the course of the operation. This is a slow method that consumes large amounts of buffer relative to sample size.

In strict isocratic elution the resolution is proportional to the square root of the column length. Therefore increasing the column length increases the resolution. This is so because both the moving velocity and the height equivalent to a theoretical plate of the sample in the column are constant during isocratic elution. Therefore, as the sample molecules move down the column, the distance between them increases. However, in gradient elution, the sample retention varies with the time, and therefore the resolution is not such a simple function of column length. For very steep gradients the resolution is independent of column length; as the gradient becomes less steep, the resolution becomes more a function of column length. Therefore to increase resolution by increasing column length requires a gradient of shallow slope.

In common gradient elution changes in the elution buffer change the affinity between the components and the matrix. This may be accomplished in the case of ion exchange chromatography, for example, by increasing or decreasing the pH or altering the ionic strength of the solution, thereby altering the affinity of sample and matrix.

Continuous gradients use a gradually changing buffer composition for elution. The buffer may vary in ionic strength, pH, or other component, and the rate of change may be linear or nonlinear with time (concave up or concave down). A step gradient is a discontinuous change from one buffer composition to another discrete level. Step gradients are more easily reproduced at large scales than are continuous gradients, an advantage that is very important in manufacturing situations. Gradient elution takes a long time and requires a large amount of elution buffers, whereas stepwise elution, bracketed around the product of interest, takes a much shorter time and is much more amenable to automated operation.

The optimal buffer compositions for the various steps, or stages, of elution may be determined through experiments involving gradient elution. That is, by monitoring the conductivity in the example of a salt gradient, as the protein elutes, one may find the concentration of the buffer that causes the protein to begin to elute and, similarly, the concentration of the buffer that causes the next impurity, contaminant, or competing protein to elute. These two compositions, then, provide the steps for eluting the desired product protein.

In the elution of proteins from an ion exchange column by stepwise gradients two types of situation arise. In the first situation sufficient elution buffer is flowed through the column so that the buffer leaving the column is identical with the buffer entering the column before the step occurs. In this case the protein moves through the column, spreading at the rate of the elution buffer. In the second case the step occurs (a new type of buffer is supplied to the column) before the previous type of buffer exits the column. Here one has at least three different buffer compositions travelling through the column at the same time, and it is possible for the front of the upstream buffer to catch and pass the protein peak as it is eluted in the downstream buffer. This may cause an artifact to appear in the chromatogram, which will show more than one protein peak and may lead one to believe that the product is not homogeneous, when in fact the peak is related to a difference in the mobile–phase composition rather than to heterogeneity of the product.

Displacement elution occurs when a substance of higher affinity for the matrix displaces a substance of lower affinity. This occurs, for example, in an ion exchange column when a large sample is applied. Proteins of less affinity will move through the column more quickly than will proteins of greater affinity, which displace the front runners from the matrix. A hierarchy of proteins in descending order of affinity is set up within the column from inlet to outlet. When proteins displace other proteins, a sharp line of demarcation between the species is rarely formed, and the adjacent peaks overlap to form regions of mixed species.

Figure 10.6 illustrates a hypothetical chromatogram of a sample containing four protein species, with the concentration of each species projected below the total at each point of the profile. To the analytical chromatographer this represents a grossly overloaded column, but to the preparative chromatographer this shows an efficient utilization of the available solid phase. A substantially greater quantity of product is obtained in pure form (the interval between the vertical dashed lines) than would be possible on this size column if displacement were not used and the concentration was brought to baseline levels between peaks. Large samples are run on preparative columns to take advantage of the inherent displacement of proteins by each other. In addition, specific molecules such as carboxymethyl dextrans may be added to the sample to displace the proteins and effect a greater separation between peaks than illustrated here. Occasionally the overloading of a column results in deteriorated performance with the later–eluting species carried forward into the earlier–eluting species with considerable overlap. This condition, known as "tag–along," probably results from the exhaustion of available sites by the early species, preventing adsorption of the later species, which then move ahead with the carrier phase.

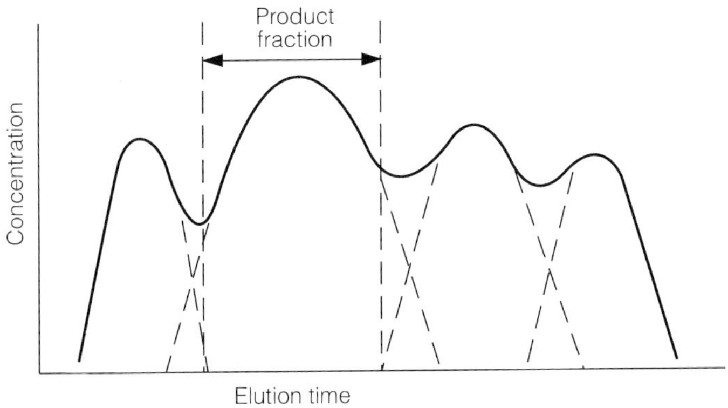

Figure 10.6. Hypothetical chromatogram, illustrating displacement by proteins in preparative chromatography with large sample size. Fraction F includes only pure protein B. Dotted lines indicate concentration of each species; solid line indicates cumulative concentration of all species.

Chromatographic operations may be automated by installation of programmed valves and pumps that deliver sample and buffer to the column while collecting fractions. Because chromatography is a batch process, the amount of sample that can be loaded on a given column is limited, and continuous operation in the classical sense cannot be achieved. However, if the sample is stable, it may be held in reservoir and applied in successive batches by means of automated equipment. Repetitive runs can thus be completed automatically, allowing the use of a small column to purify large amounts of material. This offers significant savings when an expensive solid phase is employed, such as in affinity chromatography.

10.4 EQUIPMENT

The equipment for a typical chromatographic process includes, in addition to the column containing the resin, a pump to provide sample and buffers to the column, a monitor (for ultraviolet absorbance, conductivity, or pH) to track the elution profile, and a diverter valve to allow collection of peak fractions. A typical set up is diagrammed schematically in Figure 10.7.

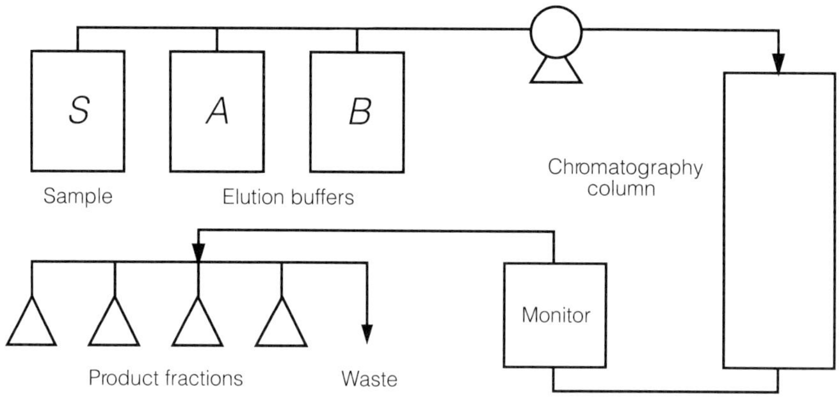

Figure 10.7. Schematic diagram of typical equipment arrangement for preparative chromatography.

The hardware that houses the solid phase also affects the resolution. Factors here include the flow distribution at the head of the column, dead spaces, piping size, and pump type. The most efficient columns are those in which the sample moves through the column at a constant velocity in a plane perpendicular to the column length, with ideal plug flow. The difficulty in preparative columns is to convert a plug of sample from narrow bore inlet pipe to a wide column. To achieve this ideal requires careful distribution of the sample to the head of the column through the use of jet deflectors at the inlet connection as well as tapered delivery and exit channels. Figure 10.8 shows the features of the inlet that must be considered to obtain maximum resolution. Dead spaces must be minimized to prevent back mixing, which dilutes the sample before loading or disperses the peak after elution.

The absence of dead space is important not only in the column but also in the pump that applies the sample and in the monitors that track the elution. Resolution is maximized when the longitudinal mixing is minimized. This requires a very sharp demarcation between the sample and elution buffers, with no mixing as they enter the column. Axial diffusion beyond the column can be minimized by keeping the lines short and their diameter narrow.

Poor arrangement and configuration of column connections and other hardware can contribute to an increased pressure drop, leading to decreased resolution. Pressure drop can be minimized by using large–diameter lines; axial mixing is decreased and plug flow is increased with small–diameter

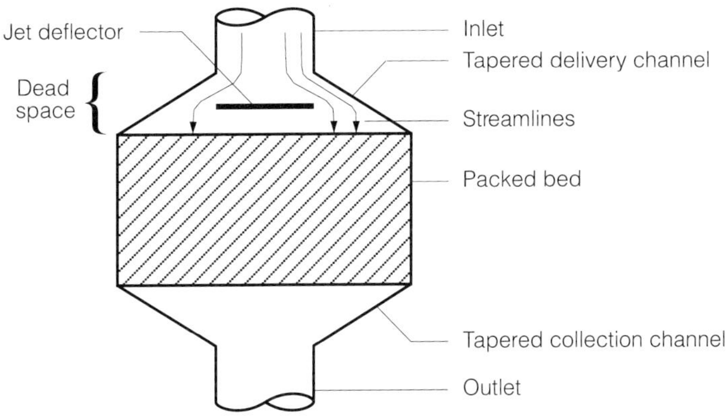

Figure 10.8. Diagram of column inlet and outlet. Dimensions of channels are exaggerated. Note the dead space and nonuniform flow, which may lead to energy loss.

lines. Obviously, some compromise must be made between pressure drop and axial mixing in selecting the optimal line diameter.

The void volume, or volume of mobile phase relative to solid phase, is slightly increased along the wall of the column, leading to slightly increased flow rates through these channels. This has less influence on the resolution of large–diameter columns than it does in smaller columns.

The uniformity of distribution of packing material through the bed has a significant impact on the resolution obtainable from any particular column. A uniform distribution of particles requires thorough mixing of the media and quick compression in the column to avoid classification by size during packing. Channels, cracks, and local density variations all reduce the resolution available in a column.

An important consideration in preparative chromatography is the materials of construction. The presence of acid or base in the process stream and chloride (which corrodes stainless steel) will have an impact on the choice of materials as well as cleaning and sanitizing procedures.

The cost of the resin is a major expense, and efforts must be expended to maintain its suitability for use. Chromatography resins for preparative separations have particle sizes of the order of 10 to 100 μm. The small channels between particles are subject to plugging if the buffers and sample are not filtered. Nonspecific adsorption of nucleic acids, lipids, and proteins will gradually build up in the column and decrease the resin capacity if not regularly removed by cleaning. Solutions containing high salt concentrations (1 to 2 M) will remove most ionic bound substances from ion exchange and gel filtration columns. Water will remove most hydrophobic bound materials. Treatment with 0.5 to 1 M NaOH for one or two hours will hydrolyze most other materials. Care should be taken with strong acid, base, and salt solutions, as some resins, column housings, and other fittings may be degraded by such contact.

Resins are subject to microbial attack, chemical reaction, and structural degradation. Many resins can be treated with hydroxide solutions to remove pyrogens and noneluted proteins. Berglof et al. (1988) found that 0.1 M NaOH was sufficient for rapid reduction of many microorganisms, including gram–negative bacteria, following a contact time of one hour. They recommend 0.5 M NaOH for one hour as a clean–in–place method for the destruction of all microbes in chromatography columns. Sofer and Nystrom (1989) report the time and temperature dependence of the destruction of various microorganisms with differeng concentrations of NaOH. They conclude that "0.1 M NaOH is an effective disinfectant for vegetative bacteria and yeast" when in contact for 1 to 2 hours. Fungal contaminations require higher concentrations, up to 1 M. Resins may be stored in preservatives such as azide, thimerosal, or 0.1 M NaOH to prevent microbial growth.

The volume of a chromatography column is limited by the structural capability of the resin medium. A column filled with soft media may result in collapse of the media and reduction of flow rates if the column is too tall. Laboratory columns are able to rely on friction between the packing material and the column wall to help support the media. Preparative columns do not have this support and must use less–compressible media or shorter column lengths. Of course, any number of chromatography columns can be joined together in series or parallel to provide long length and large capacity.

The pressure one selects for operation is governed by several factors. The advantage of high pressures is twofold: speed of operation and small–particle–size (higher–resolution) media. The disadvantages of high operating pressures include limited selection of media, since popular compressible media are not suitable; the higher cost of high–pressure equipment, which must be made of high–strength metals, such as stainless steel or titanium; and the high cost of small particle–size media. High–pressure liquid chromatography, also known as high–performance liquid chromatography (HPLC), because of the high number of theoretical plates relative to low–pressure (atmospheric–pressure) liquid chromatography, has become popular for preparative chromatography of proteins because of the high resolution, particularly in reversed–phase operations.

10.5 SCALE UP

Several aspects of scale up are universal to all chromatographic techniques. The goal of scale up is to reproduce the small–scale process, including resolution, time, and yield at an increased volume. The process should be fully optimized with respect to these three variables at the small or laboratory scale. The matrix material, column length, linear flow rate (volumetric flow rate divided by cross–sectional area), sample concentration and sample volume should be based on laboratory–scale optimization studies. These parameters do not change upon scale up.

Volumetric throughput is related to column volume, which is increased to give a column volume proportional to the increased sample volume. Nearly all columns are cylindrical with flow parallel to the axis. Scaling up a column means increasing the diameter to give a volume proportional to the increased sample volume. For example, if a laboratory column 2.5 cm in diameter and 30 cm long that has been optimized to handle a 1–g sample needs to be scaled up to handle a 100–g sample, the preparative column would have 100 times the volume or 10 (the square root of 100) times the diameter, or a diameter of 25 cm. The volumetric flow rate increases by a hundredfold, but the linear flow rate and all other operating parameters remain the same.

The aspect ratio, or height–to–diameter ratio, affects the length of time required for a chromatographic operation, which in turn affects the cost of the process step. In gel filtration chromatography the column length is set by the resolution requirement and the diameter by the batch volume. For other processes the length of the column is not so clearly defined. Janson and Hedman (1982) recommend aspect ratios of 0.37 for ion exchange and 0.55 for desalting. The difficulty of constructing a large–volume column with low aspect ratio is somewhat reduced by the axial–flow design column, which packs the resin in a cylindrical annulus with the flow perpendicular to the axis of the cylinder.

Analytical elution chromatography is characterized by linear isotherms. It is possible to scale up an analytical run by keeping a small sample–to–column–volume ratio, but it is a more cost–effective use of the column and process time to switch to the displacement mode, in which the isotherms are nonlinear and the sample–to–column–volume ratio is high. The displacement mode can be operated on the small scale to optimize it and then scaled up. Helfferich (1986) has suggested that, if analytical developments had not been so successful over the years, the efforts to scale them up without switching to the displacement mode might have been avoided, to the advancement of the preparative chromatography industry.

REFERENCES

Berglof, J. H., N. P. Adner, and S. Y. Doversten, "Inactivation of microbial contamination in chromatographic separation media using sodium hydroxide," paper presented at the XX Congress of the International Society of Blood Transfusion, London, July 10–15 (1988).

Gariel, P. and R. Rosset, "La chromatographie en phase liquide preparative par developpement par elution," Analusis 10(9):397–408 (1982).

Giddings, J. C., Dynamics of Chromatography, Marcell Dekker, New York (1965).

Giddings, J. C., and H. Eyring, "A molecular dynamic theory of chromatography," *Journal of Physical Chemistry* 59:416 (1955).

Helfferich, F. G., "Theory of multicomponent chromatography," *Journal of Chromatography* 373:45–60 (1986).

Janson, J. C., and P. Hedman, "Largescale chromatography of proteins," *Advances in Biochemical Engineering* 25:43–99 (1982).

Kopaciewicz, W., S. Fulton, and S. Y. Lee, "Influence of pore and particle size on the frontal uptake of proteins," *Journal of Chromatography* 409:111–124 (1987).

Knox, J. H., and H. M. Pyper, "Framework for maximizing throughput in preparative liquid chromatography," *Journal of Chromatography* 363:1–30 (1986).

Regnier, F. E., and K. M. Gooding, "High performance liquid chromatography of proteins," *Analytical Biochemistry* 103:1–25 (1980).

Sofer, G. K., and L.–E. Nystrom (eds.), Process Chromatography: A Practical Guide, Academic Press, San Diego (1989).

Van Deemter, J. J., F. J. Zuiderweg, and A. Klinkenberg, "Longitudinal diffusion and resistance to mass transfer as causes of nonideality in chromatography," *Chemical Engineering Science* 5:271–289 (1956).

Vanecek, G., and F. E. Regnier, "Variables in the high–performance anion exchange chromatography of proteins," *Analytical Biochemistry* 109:345–353 (1980).

Vermeulen, T., M. D. LeVan, N. K. Hiester, and G. Klein, "Adsorption and Ion Exchange," in Chemical Engineers' Handbook, 6th ed., R. H. Perry, D. W. Green, and J. A. Maloney (eds.), McGraw–Hill, New York (1984).

Yang, C. M., and G. T. Tsao, "Packed–bed adsorption theories and the applications to affinity chromatography," *Advances in Biochemical Engineering* 25:1 (1982).

FURTHER READING

Fasold, H., "Chromatography of proteins," in Chromatography, E. Heftmann (ed.), Van Nostrand Reinhold, New York (1975).

Grubner, O., "Statistical moments theory of gas–solid chromatography: diffusion–controlled kinetics," in Advances in Chromatography, vol. 6, J. D. Giddings and R. A. Keller (eds.), Marcel Dekker, New York, pp. 173–209 (1968).

Grushka, E., "Chromatographic peak shapes: Their origin and dependence on the experimental parameters," *Journal of Physical Chemistry* 76(18):2586–2593 (1972).

Helfferich, F., and G. Klein, Multicomponent Chromatography, Marcel Dekker, New York (1970).

LeVan, M. D., and T. Vermeulen, "Effects of channeling and corrective radial diffusion in fixed–bed adsorption using activated carbon," in Fundamentals of Adsorption, A. L. Myers and G. Belfort (eds.), Engineering Foundation, New York (1983).

Regnier, F. E., "Chromatography of complex protein mixtures," *Journal of Chromatography* 418:115–143 (1987).

APPENDIX 10 Vendors of preparative chromatography equipment.

Amicon 24 Cherry Hill Dr. Danvers, MA 01923 800-343-1397	Columns, media, high pressure equipment
DuPont Barley Mill Plaza Wilmington, DE 19898 800-453-8869	Media
IBF Biotechnics 8510 Corridor Road Savage, MD 20763 301-498-0511	Media
ISCO Box 5347 Lincoln, NE 68505 800-228-4250	Monitors, collectors
Kontes 1022 Spruce St. Vineland, NJ 08360 609-692-8500	Columns, media
Perstorp Biolytica AB S-223 70 Lund Sweden 011-4646-168780	Media

Pharmacia LKB Biotechnology Inc. Centennial Ave. Piscataway, NJ 08854 800-922-0318	Columns, media, 800 monitors, collectors
Prochrom 5622 West 73rd Street Indianapolis, IN 46278 317-297-1667	Media, high–pressure equipment
SepTech, Inc. 2 Columbia St. equipment Wakefield, RI 02879 401-789-5660	Media, high–pressure
The PQ Corporation 280 Ceder Grove Road Lafayette Hill, PA 19444 215-941-2074	Media
TosoHaas 82 Cummings Park Woburn, MA 01801 617-932-4712	Media
Waters Chromatography Division Millipore Corporation 34 Maple St. Milford, MA 01757 508-478-2000	Media, high–pressure equipment

CHAPTER 11

ION EXCHANGE CHROMATOGRAPHY

Ion exchange chromatography is a separation process based on the relative affinities of various sample components for ionic sites on a solid matrix. Ion exchange is a versatile separation method, with applications throughout the chemical–process industries. Following the discovery by Adams and Holmes (1935) that crushed phonograph records exhibit ion exchange properties, the use of organic resins as the solid–phase matrix has largely supplanted earlier applications of natural clays. The wide variation in properties afforded by organic resins has led to many diverse uses, from early work by Tiselius in protein purification to cheese manufacture, water purification, and electroplate–solution recovery.

Among the tools available for the separation of proteins, ion exchange is one of the most valuable. The amphoteric nature of proteins ensures that conditions exist under which almost any soluble protein can be purified by ion exchange. And modern resins, developed specifically for protein work, avoid denaturation of sensitive biological molecules, a problem with some early ion exchangers that came from the water–purification field. While the protein must be soluble, ion exchange resins will withstand harsh conditions, allowing the use of nonionic detergents and chaotropic agents to maintain sample solubility.

Several phrases and words are commonly used interchangeably in the literature to describe the solid phase in ion exchange chromatography. These include "ion exchanger," "solid phase," "matrix," "resin," and "medium" (or "media"). Though not all ion exchangers are made from organic–polymer resins, most currently available ones are. Indeed, as Helfferich (1962) points out, it is the systematic study of organic resins that has lead to the current diversity and widespread application of ion exchange as a separation process.

11.1 MECHANISM

Proteins are made up of amino acids that contain various chemical groups attached to a peptide backbone (see Chapter 3). These groups may be positively or negatively charged, or they may be electrically neutral. The ion exchange resin (solid support matrix) contains electrically charged species, such as carboxyl or quaternary ammonium groups, covalently attached. Associated with each linked charge is a counterion of opposite charge, bound only by ionic attraction.

Charged regions of the protein containing a charge of opposite polarity to the resin will be attracted to the solid phase. The charged site on the protein will displace the counterion, and become bound to the resin by ionic interaction. Noncharged species will not become attached to the resin and may be separated by washing. The protein bound by ionic interaction can be removed or eluted from the resin by displacement with an ion having a stronger affinity for the solid phase than the protein does. Figure 11.1 illustrates the basic idea. The ions that are exchanged are the counterions associated with the fixed–charge groups of the resin, and displacement occurs in the order of relative affinity. The solid phase does not undergo alteration during the process, rather one counterion is exchanged for another, with the ion having the greater affinity for the resin displacing the one with a lesser affinity.

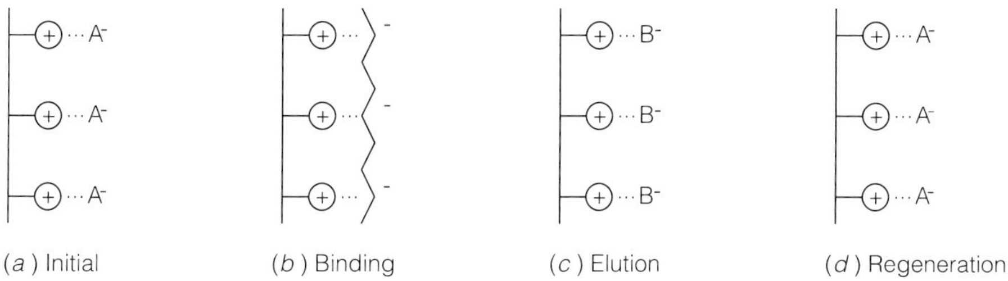

Figure 11.1. Schematic diagram illustrating mechanism of ion exchange. (*a*) Anion exchange resin with negatively charged counterions. (*b*) Negatively charged protein associated with resin. (*c*) Elution of protein from resin. (*d*) Regeneration of resin to original counter ion.

11.1.1 Ionic Nature of Proteins

Proteins carry both positively and negatively charged groups and are called "amphoteric." The charge they carry is dependent upon the pH, with the pH value at which they have zero net charge termed the "isoelectric point." To determine the optimal conditions for operating an ion exchange separation, we shall find it helpful to know the variation in charge of the protein with pH and the isoelectric point. It is also helpful to know the pH range in which the protein is stable, as this defines the working range for ion exchange separations.

Table 11.1. Conditions under which a protein binds or does not bind to the ion exchange matrix.

	Cation exchanger	Anion exchanger	Protein net charge
Above pI	Does not bind	Binds	Negative
Below pI	Binds	Does not bind	Positive

At a pH below the isoelectric point the net charge of the protein is positive, and the protein will bind to a cation exchanger; at a pH above the isoelectric point the net charge of the protein is negative, and the protein will bind to an anion exchanger. Table 11.1 describes the conditions under which a protein does or does not adsorb to the matrix. The appropriate exchange matrix must be chosen on the basis of the stability of the protein and the net charge of the protein in that range.

Although the net charge of a molecule may be zero at the isoelectric point, the local charge distribution is not uniform, and areas of positive or negative charge will be found across the surface of the protein. Therefore, whereas the isoelectric point may serve as a guide for selecting conditions for ion exchange, it will not always predict the point at which the molecule no longer has an affinity for the matrix.

11.1.2 Isotherms and the Law of Mass Action

When a solution containing ionic species is brought into contact with an ion exchange matrix, a dynamic equilibrium is established as the ionic species

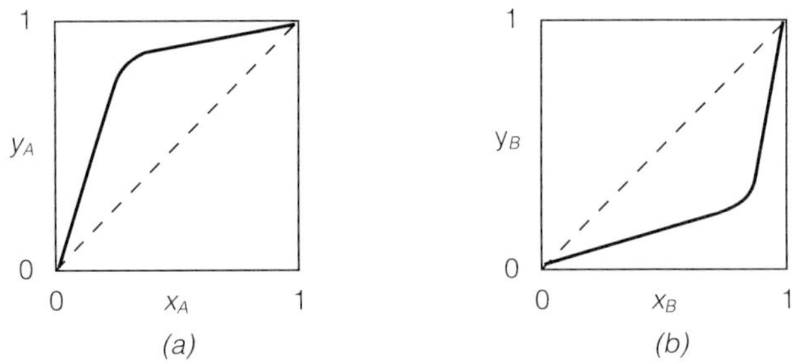

Figure 11.2. Ion exchange isotherms for a two-component system, where y is the relative concentration of one component in the solid phase and x is the relative concentration of the same component in the liquid phase. (*a*) Component *A* has greater affinity for the resin than does component *B*. (*b*) The isotherm for *B* is the mirror image for *A* (along the diagonal) since $x_A + x_B = 1$. Each additional component adds an additional dimension to the isotherm.

distribute between the phases. The distribution of a particular moiety between the solid and liquid phases depends on the concentration of that species. When all conditions, including temperature, are held constant and the relative concentration of exchangeable ions is varied, the result may be plotted as an ion exchange isotherm. Figure 11.2 shows an isotherm for a two–component system, such as protein *A* and salt *B*.

The species with the greater affinity for the resin will have a relative concentration in the solid phase greater than the relative concentration in the liquid phase ($y > x$), while the species with the lesser affinity for the resin will have a greater concentration in the liquid phase than in the solid phase ($y < x$). Because the distribution of a species between the solid and liquid phases varies with temperature, isotherms reflect the phase distribution only for a single temperature.

The distribution of a species between solid and liquid phases at equilibrium is described by the law of mass action. As outlined by Vermeulen et al. (1984) the exchange reaction can be written as

$$mA + RmB = mRA + B \tag{11.1}$$

where k is the valence ratio of ions A and B, and R is the solid phase. The general equilibrium equation may be represented by

$$K = \left(\frac{q_{\mathrm{A}}}{c_{\mathrm{A}}}\right)^{\mathrm{k}} \left(\frac{c_{\mathrm{B}}}{q_{\mathrm{B}}}\right) \tag{11.2}$$

where K is the distribution coefficient, and q and c are concentrations of charge species in the solid and fluid phases, respectively. If we replace q with y/Q_0, where Q_0 is the total number of charge sites on the resin (y is the relative concentration of charge sites in the solid phase), and replace c with x/C_0, where C_0 is the total ionic strength of the solution (x is the relative concentration of charges in the fluid phase), the equation becomes

$$K = \left(\frac{C_{\mathrm{o}}}{Q_{\mathrm{o}}}\right)^{\mathrm{k}-1} = \left(\frac{y_{\mathrm{A}}}{x_{\mathrm{A}}}\right)^{\mathrm{k}} \left(\frac{x_{\mathrm{B}}}{y_{\mathrm{B}}}\right) \tag{11.3}$$

This is the equation for the generalized isotherm shown in Figure 11.2. For a simple salt ion exchange, such as sodium's replacing calcium, $k = 2$. For proteins, k will be some integer or fraction greater than 1. For example, consider the case where $k = 2$. The equation reduces to

$$K = \left(\frac{C_{\mathrm{o}}}{Q_{\mathrm{o}}}\right) = \left(\frac{y_{\mathrm{A}}^2}{y_{\mathrm{B}}}\right) \left(\frac{x_{\mathrm{B}}}{x_{\mathrm{A}}^2}\right) \tag{11.4}$$

At low ionic strength ($KC_0 << Q_0$), species A will be preferred by the resin, and counterion B will be displaced from the resin by A. At high ionic strength ($KC_0 >> Q_0$), the equilibrium is reversed, and counterion B will displace A. Thus the affinity between an ion and the matrix is a function of the concentration. In general, at low ionic strengths, ions with greater numbers of charged sites have higher affinity for the resin than do ions with fewer charge sites. The reversal in this behavior that occurs at high ionic strengths is what makes regeneration possible, makes ion exchange a reversible process, allows the resin to be reused, and greatly reduces the cost of operation. For example, Yamamoto et al. (1988) measured the distribution coefficient of beta–lactoglobulin A on a DEAE ion exchange matrix at pH 7.7 and 20°C and found a decrease in K from 500 to 1 with an increase in ionic strength from 0.1 to 0.3.

11.2 MATRICES

The solid–phase matrix exerts considerable influence over the ion exchange operation, affecting not only the resolution of the step but also the cost and ease of use. Several different matrices are available, as shown in Table 11.2. Those in most common use contain a hydrophilic backbone, which avoids problems of denaturation of the protein, as may occur with hydrophobic resins.

The capacity of the resin is measured as both the total number of charges per unit volume or mass (for example, milliequivalents per milliliter) and as mass of protein that can be adsorbed per unit of volume (for example, milligrams per milliliter). The active capacity, or the mass of protein that the resin can adsorb, depends on the number and size of the pores in the resin particle. Larger pores make the charge sites more accessible to the protein, while small pores may exclude much of the total charge because small pore sizes hinder the diffusion of proteins and binding sites within these small pores are not available to large molecules.

The pore size is usually described as an exclusion limit, that is, the largest globular protein (measured in daltons) that can penetrate the particle. The distribution of pore sizes is an important parameter in estimating the capacity of an ion exchange matrix for a particular protein, but this information is rarely available in the literature.

The capacity of an ion exchanger, in terms of total charge present, is determined by titration with a strong acid or base. For any given protein the capacity, in terms of protein binding, can be accurately determined only by empirical measurement.

The carbon backbone of many matrices renders them vulnerable to microbial

Table 11.2. Limited comparison of typical ion exchange media.

Polymer base	Brand name	Capacity (meq/ml)	Particle size (μm)	Exclusion (D)
Agarose	Sepharose	0.1	100	4,000,000
Cellulose	Sephacel	0.1	100	1,000,000
Dextran	Sephadex	2	100	200,000
Ethylene glycol methacrylate	Toyopearl	0.1	100	5,000,000

attack. Some resins are alkali resistant and may be washed with sodium hydroxide to aid preservation, while others require careful storage and use. Other resins, particularly those based on silica, are subject to degradation at high pH.

The carbon backbone, which makes resins nondenaturing and makes them subject to microbial degradation, tends to be soft in structure and requires special consideration for column design and packing. Some resins contain considerable cross–linking agents to provide structural support; others tend to be spongy gels. The less–firm matrices are limited in terms of flow rate because of the compression of the column that may occur. Figure 11.3 shows the pressure as a function of flow rate for an agarose–based medium. As the flow rate increases, the pressure drop goes up at an increasing rate, indicating that the medium is being compressed.

A specialty matrix that finds occasional use is hydroxyapatite, $Ca_5(PO_4)_3OH$, which appears to function as an anion exchanger of low

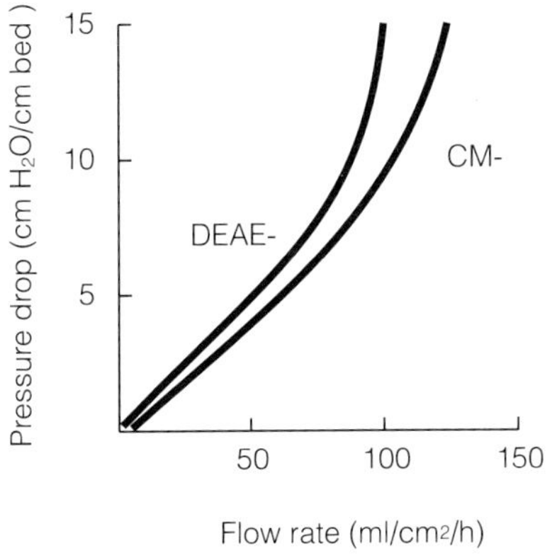

Figure 11.3. Pressure drop as a function of flow rate for DEAE- and CM-Sepahrose CL-6B. (Courtesy of Pharmacia LKB Biotechnology.)

charge density, with high molecular weight proteins and nucleic acid binding preferentially over smaller molecules. As described by Karlsson et al. (1989), the drawback of mechanical instability of hydroxyapatite has been overcome to a degree by ceramic fluoroapatite.

11.2.1 Shrinking and Swelling

Many ion exchange matrices experience significant volume changes with changes in ionic strength or counterion type. These volume changes result from a change in the activity of the molecules inside the resin particles, which attract or repel solvent molecules and cause the particle size to alter. The extent of size variation is limited by the crosslinking between polymers.

Such shrinking and swelling is usually apparent only at very high salt concentrations and for most protein work with rigid media is not a critical consideration; but there are times when a soft matrix may be used, and to someone who has never seen its shrinking and swelling before they are dramatic. It is also important that, if the bed is packed with the resin in the shrunk form, sufficient head room be available when it starts to expand; otherwise the column may crack. Variations in swelling and shrinking will be more pronounced if a wide range of ionic strength is encountered rather than a narrow range, particularly if one goes from a high–ionic–strength to a very–low–ionic–strength buffer.

11.2.2 Strong and Weak Exchangers

The difference between the so–called weak ion exchangers and strong ion exchangers is the pH range over which they retain their charge. Just as acetic acid is ionized only at a pH above 4, a weak cation exchange resin is ionized only at a pH above its dissociation constant. Strong ion exchangers, on the other hand, remain dissociated over a very wide range, just as strong acids such as sulfuric acid are ionized over a wide pH range. Such weak cation exchangers as carboxymethyl groups tend to lose their charge at pH values below 6, and weak anion exchangers, such as DEAE, tend to lose their charge above pH 9. Therefore, if the desired working range is outside these ranges, then strong ion exchangers should be used. Otherwise it makes no difference whether a strong or weak exchanger is used in terms of charge availability although weak ion exchangers may have less tendency to denature labile proteins than strong ion exchangers do. Figure 11.4 shows titration curves of weak and strong exchangers.

Weak ion exchangers may have a decreased capacity in a high–ionic–strength buffer; so an ion exchange resin that has high capacity,

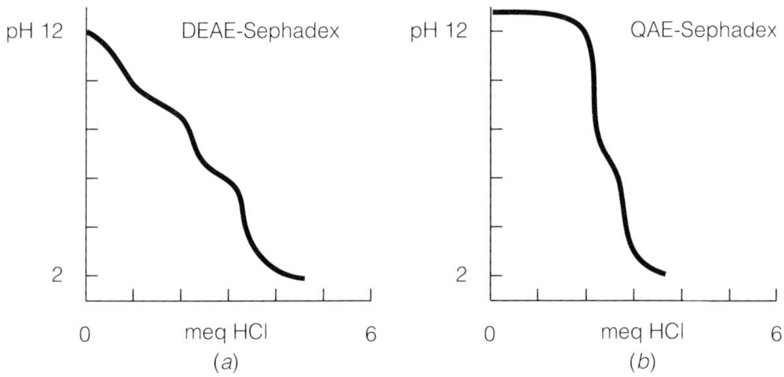

Figure 11.4. Titration curves for (*a*) weak and (*b*) strong anion exchangers (1 g resin in 1 M KCl). (Courtesy of Pharmacia LKB Biotechnology.)

even in high–ionic–strength buffers, should be chosen for proteins that are unstable in low–ionic–strength buffers. Weak ion exchangers, which have a tendency to be titrated by the buffer, are a poor choice for a buffer gradient elution system as the interface (between the pH of the upstream buffer and the pH of the downstream buffer) will show a variation in pH intermediate to preceding and following positions in the column. It is important to note that the terms strong and weak do not refer to the binding affinity or the stability of the matrix; they refer only to the extent of dissociation in the classical sense of acid–base chemistry.

Chromatofocusing is a form of ion exchange chromatography in which a resin is equilibrated at one pH and eluted at a second pH. The use of a weak ion exchange resin causes a pH gradient to be formed at the solvent front owing to the buffering action of the resin. This pH gradient in turn leads to an ordering of proteins by isoelectric point. Molecules of charge sign opposite the resin bind; those of charge sign like the resin do not bind. For best resolution the equilibration–buffer pH and the elution–buffer pH should bracket the pI of the product protein. For example, Hutchens (1989) describes the elution of a protein of pI 9 from a weak anion exchanger equilibrated at pH 8 by elution with buffer at pH 10. One difficulty often encountered in chromatofocusing is the tendency of many proteins to precipitate at their isoelectric points, especially in the low ionic strength buffers that are commonly used.

11.2.3 Charge Groups

A wide variety of both strong and weak charge groups attached to various matrices are sold commercially. Some of the more common ligands are listed

Table 11.3. Comparison of ion exchange charge groups.

Type	Abbreviation	Name	Structure
Strong cation	S	Sulphonate	$-CH_2SO_3^-$
Strong cation	SP	Sulphapropyl	$-CH_2CH_2CH_2SO_3^-$
Weak cation	CM	Carboxymethyl	$-CH_2COO^-$
Weak cation	P	Phospho	$-PO_4H_2^-$
Strong anion	Q	Quaternary amine	$-CH_2N^+(CH_3)_3$
Strong anion	QAE	Quaternary aminoethyl	CH_2CH_3 $-CH_2CH_2N^+CH_2(OH)CH_3$ CH_2CH_3
Weak anion	DEAE	Diethyl aminoethyl	CH_2CH_3 $-CH_2CH_2N^+H$ CH_2CH_3

in Table 11.3. Though not all charge groups are available on every matrix, the process designer is offered a wide selection.

11.2.4 Capacity

The capacity of an ion exchanger is important both in terms of scale up and in terms of cost of the operation. We can determine the capacity by operating a small column under the same initial conditions we expect in the final operation and tracing the product as it leaves the column when it first starts to appear in the column effluent. Then we have an idea of the capacity of the column, that is, the amount of material applied to the column at the time that the product first starts to break through the end of the column and is no longer binding because sites are no longer available in the column.

An alternative method is to apply the product of known concentration of starting buffer in a batch method to the resin, stir, and then measure the amount of product remaining in the liquid. Either method provides roughly the same answer for total column capacity.

11.3 OPERATION AND APPLICATION

Ion exchange is a batch operation consisting of three steps: sample loading, sample elution, and solid–phase regeneration. In the first step the multivalent protein is adsorbed to the solid phase, and the counterion is displaced and moves into the fluid phase. For this to occur the adsorbing protein must have a greater affinity for the resin than does the existing counterion; that is, the isotherm must be convex upward.

In the second step an eluant is applied to displace the protein from the solid phase. The eluant in ion exchange reduces the affinity of the protein for the resin relative to ions in the buffer in one of two ways: either by increasing the ionic strength of the buffer (by increasing the salt concentration) or by changing the charge of the protein (by altering the pH). In both cases the net effect is a change in the isotherm from convex upward to convex downward, or from Figure 11.2(*a*) to 11.2(*b*).

A third alternative is to displace the protein by another protein or similar multivalent polymer. "Displacement chromatography," as this method is called, has the advantage of high resolution with much smaller columns than is required for conventional elution—a factor that may very well control the use of ion exchange for the production of large–volume and low–value proteins. In some cases large sample loads will set up a natural displacement between the proteins in the sample and provide a higher degree of separation than a small sample load can. Alternatively a displacing ion can be added to the sample to effect the separation. This method has the disadvantage of requiring subsequent removal of the displacing polymer. Displacement chromatography is reviewed in detail by Torres et al. (1984, 1979) and Horvath et al. (1981).

The third step is regeneration of the ion exchanger by stripping off all remaining protein and returning the resin to its original counterion form. Elution with a high–ionic–strength buffer (such as 1 M NaCl) may accomplish both objectives. Alternatively the protein may be hydrolyzed with alkali (such as 0.5 M NaOH), after which a high ionic strength solution restores the counterion.

11.3.1 Batch Mode

Ion exchange may be run in either a batch or a column mode. Batch separations are simple and easy to conduct although they require multiple manipulations of the gel that can lead to its mechanical breakdown. Although both adsorption and elution may be carried out in batch fashion, following batch adsorption the resin is commonly packed into a column and then eluted. Batch elution is equivalent to a single theoretical plate, compared to several hundred in a column. Even so, batch operations may be practical, especially for crude steps at the beginning of the process.

11.3.2 Elution

The purpose of the elution step is to remove the protein of interest from the ion exchange resin with high resolution, that is, distinctly separated from adjacent components. Of the three types of elution, isocratic, continuous gradient, and step gradient, the first is rarely used in ion exchange, the second is common in development applications, and the third is most common in large–scale processes.

Continuous pH gradients are difficult to reproduce and should be avoided in large–scale production. The difficulty with pH gradients is that, as the pH changes, the ionic strength varies also; therefore the ionic strength cannot be kept constant. Linear pH gradients cannot be easily formed, and ion exchangers (particularly weak exchangers) have a high buffering capacity that reacts with the pH gradient as it is applied, titrates the buffer, and results in a change in the pH through the length of the column. Stepwise pH gradients are much simpler to use than are continuous pH gradients, but even in a stepwise gradient the buffer must titrate the ion exchanger. If the counterion on the resin is hydroxide or hydrogen ion, then an increase in the salt concentration of the eluting buffer will tend to displace the counterions and lead to a pH change.

Continuous ionic–strength gradients are simple to reproduce. Unlike pH gradients, the combination of two different buffers of differing ionic strength is additive, and each contributes proportionately to the final ionic strength of the solution. The length of the gradient will affect the resolution. Longer gradients lead to band broadening, while shorter gradients may not give sufficient time for equilibration between the buffer front and the ion exchanger, with poor resolution resulting. The total volume of eluant in a gradient should be four to five times the bed volume.

Stepwise ionic–strength gradients are simple to use and provide the greatest reproducibility in the manufacturing environment. Figure 11.5 shows the typical application of a step gradient. The ionic strength at which the product protein begins to elute is determined by the use of a continuous gradient. Buffer having an ionic strength slightly lower than necessary for product elution is applied until no further elution occurs; the next buffer applied is the proper ionic strength to elute the product; this is followed with a high–ionic–strength buffer to strip the remaining protein and regenerate the column. One disadvantage of step gradients is that substances that tend to elute close together in a continuous gradient may be undifferentiated in a step gradient, resulting in more overlap than that obtained with a continuous gradient.

The best resolution is obtained when the protein of interest is bracketed between adsorption and elution as closely as possible. This means that at

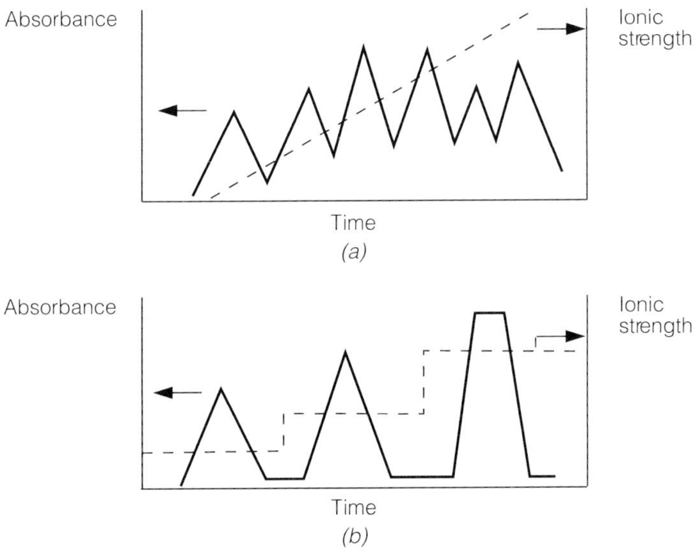

Figure 11.5. Comparative profiles of (*a*) continuous and (*b*) step gradients. The solid line represents the absorbance, and the dashed line conductivity.

the time of adsorption the product protein should have the lowest affinity for the resin of all the proteins that adsorb to the resin, and at the time of elution the product protein should be the first to desorb. Once can accomplish this by binding the protein under conditions where the ionic strength is as high as possible and the material still binds and by eluting the product with an ionic strength that is as low as possible and still permits the product to elute. Under these conditions, the least impurities will bind during the adsorption step and the least impurities (of those bound) will be eluted during the elution step. Bracketing the binding and elution in this manner will simplify the operation and provide the best resolution. The stabilizing effect of salt on proteins must also be taken into account, and therefore the minimum ionic strength may be governed by the protein solubility rather than exclusively by the preferred elution conditions.

11.3.3 Additional Considerations for Resolution

Economy of operation dictates that the flow should be run as fast as possible while maintaining the resolution. But resolution to a limited extent is a function of temperature and tends to be higher at room temperature than it would be when cold, owing to mass–transport limitations arising from increased viscosity and decreased diffusivity as the temperature decreases.

This is especially pertinent for large molecules such as proteins. The ionic strength is also a function of temperature, decreasing as the solution becomes colder. With regard to temperature Yamamoto et al. (1988) suggest that, except for very unstable molecules, rapid separation at room temperature is usually superior to operation at a reduced temperature.

The flow rate divided by the cross–sectional area of the column equals the linear velocity; the range for ion exchangers of 100–μm particle diameter is 50 to 100 ml/cm^2 h. For the large–scale column the linear flow rate obtained for the pilot work should be duplicated. Therefore the total flow is increased, but the time it takes for the material to travel from the top of the bed to the bottom of the bed is the same since the large–scale column has the same length as the pilot–scale column.

11.4 OPTIMIZATION AND SCALE UP

This section covers the questions that need to be answered and the experiments that need to be run to optimize and scale up an ion exchange operation. Once the step has been optimized at small scale, the equipment can be scaled up. The choices available to the process designer are listed in

Table 11.4. Alternative choices in selection of ion exchange chromatography step parameters.

Alternative	Limiting factor
1. Batch or column	Design choice
2. Product adsorbed or not	Design choice
3. Medium selection Cation or anion Strong or weak	Design choice
4. Matrix Ease of cleaning Ruggedness Compressibility Capacity	Cost, availability, regulations
5. Particle size Low pressure, 0–1 atm, 60–100 μm Medium pressure, 1–2 atm, 30–50 μm High pressure, 10–20 atm, 5–10 μm	Cost, requirement for resolution
6. Buffer pH Ionic strength Type of gradient	Solubility, production control

Table 11.4. The selection of a particular alternative may be limited by specific considerations such as requirements for resolution, availability of materials, or external regulations; or the selection may be limited by factors involved in the overall design and be more at the discretion of the designer.

11.4.1 Mode of Operation

We must first decide whether we want to operate in batch or column mode and whether or not we want to adsorb our product to the resin. The first point depends mainly on the position of this ion exchange step in the overall process and how much resolution we require. Batch adsorption is most suitable early in the process as a recovery step, where high resolution is not critical. Generally speaking, better resolution is obtained when the product is adsorbed to the resin than when it is washed through without binding. It is simpler to get a single product protein to bind to the resin than to find conditions under which all the contaminants will adsorb. However, the sample may contain proteases or other substances that degrade our product; or the product may be very labile, and a quick flow–through step may be preferred.

11.4.2 Resin–Buffer–Elution Scheme

The choice of the resin–buffer–elution scheme is a complicated decision that requires systematic experimentation for arrival at the optimal conditions. The nature of the sample will narrow the available choices, as will the batch volume and flow rate.

The volume impacts the cost by dictating the amount of resin needed, which depends in turn on the resin capacity and on the batch size. Flow rate translates to economy, that is, faster processes cost less. This means that the flow capacity of the ion exchanger and the ability of the exchanger to react quickly with the material being adsorbed is an important consideration. But high flow rates require a rigid resin, limiting the choice of matrices. Resolution is also dependent on flow rate, and different ion exchange matrices will give different degrees of resolution, even under similar conditions. As described by Maskalick and Abbott (1987), the flow rate is also limited by the rate at which the protein binds, a trait that varies with resin and charge conditions. Whether the exchanger can be reused is something that must be decided in terms of the cost of the resin and the value of the product.

The nature of the product to be purified and its associated impurities will dictate the choice between cation and anion or strong and weak charge groups, as well as the buffer and elution conditions. The isoelectric point (pI) of the protein determines the appropriate charge and pH for binding to the resin. As shown in Table 11.1, above the pI point the protein binds to anion exchangers; below the pI point the protein binds to cation exchangers. Unless the contaminants are well characterized, including knowledge of each pI (which is rarely the case), the choice of resin is best made by comparison of three or four different matrices. The preferred matrix will bind the fewest contaminants along with the product.

Substances in the buffer serve two functions: The first is regulation and maintenance of constant pH during the operation; second, the ions in the buffer displace the protein or are displaced by the protein during the ion exchange reaction. The buffer ion should have a charge of the same sign as the fixed group on the resin (that is, a negative charge for a cation exchanger) to avoid binding of the buffer agent to the resin.

The optimal buffer conditions must be determined empirically. Generally the starting pH is one pH unit away from the pI of the product. Data of Lampson and Tytell (1965) indicate that substances tend to dissociate from ion exchangers at about 0.5 pH units from their isoelectric points at an ionic strength of 0.1 M.

The initial pH may be selected by the method listed in Table 11.5 and shown in Figure 11.6. A small–scale experiment can determine the amount of protein adsorbed by the resin at each pH. The preferred pH will be that at which the protein is bound but where it can be easily desorbed. At a less optimal pH the protein may be difficult to desorb. The distribution coefficient K is a strong function of the net charge on the protein and, consequently, a strong function of pH.

Table 11.5. Selection of starting pH.*

1. Set up series of 10 tubes, 15 ml each.
2. Add equal amount of ion exchanger to each (1–2 ml).
3. Equilibrate each tube by 10 washings of buffer (0.5 M).
   anion range = pH 5–9
   cation range = pH 4–8
4. Equilibrate by 5 washings of buffer (0.05 M).
5. Add equal amount of sample to each tube.
6. Mix tubes for 5–10 minutes.
7. Allow gel to settle.
8. Assay supernatant.

* From Anonymous (1983).

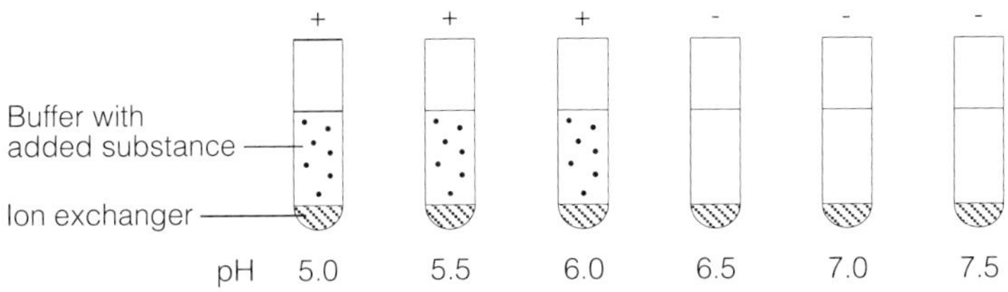

Figure 11.6. Simple test for selection of initial pH. (Courtesy of Pharmacia LKB Biotechnology.)

The pilot study should include the following steps:

1. Measure the pI of the product.
2. Select three or four alternate resins.
3. Pick the starting buffer.
4. Select resin and buffer.
5. Elute with continuous gradient.
6. Determine bracketing–step gradient.
7. Measure resin capacity.
8. Determine column length.
9. Verify small–scale trial.
10. Determine volume of full–scale column.

11.4.3 Aspect Ratio

The column aspect ratio is the length of column divided by the width. The width of the column is determined by the volume of the sample that must be applied. The height of the column is determined by the adsorption characteristics of the ion exchanger. The 10 to 30 percent of the bed where the initial binding occurs should be the upper portion of the bed; a convenient rule of thumb is that the volume of the bed should be about five times the minimum capacity, as determined by breakthrough capacity measurements.

Ion exchange columns on the small scale are typically about 20 cm long. This provides an aspect ratio, for, say, a column 5 cm diameter of four. When an ion exchange column is scaled up, the diameter is increased to handle an increased volume of sample. Once the length has been determined, it may be kept the same throughout the scale up. Thus a 20–cm bed height may be maintained, even as the volume increases and the diameter increases, to give an aspect ratio of less than one: that is, a diameter greater than 20 cm, perhaps two or three times that diameter.

11.4.4 Scale up

The scale up of an ion exchange column is straightforward. The resolution and yield at this point have been determined with small–scale columns: The conditions have been set and optimized. All that is necessary to increase the scale is to increase the volume of the column. The length of the column has also been set; so increasing the volume means increasing the column diameter and the flow rate. The linear velocity of the fluid phase (flow rate divided by cross–sectional area) should be the same at large scale as at small scale. If all operating parameters are maintained constant, the performance at full scale should be nearly identical with the performance at small scale. Some adjustments may be required due to differences in the actual equipment available.

REFERENCES

Adams, B. A., and E. L. Holmes, *Journal of the Society of Chemical Industries* (London) 54:1T (1935), referenced in Helfferich (1962).

Anonymous, "Ion exchange chromatography: principles and methods," Pharmacia, Uppsala (1983).

Helfferich, F., Ion Exchange, McGraw–Hill, New York (1962).

Horvath, C., A. Nahum and J. H. Fren, "High performance displacement chromatography," *Journal of Chromatography* 218:365–393 (1981).

Hutchens, T. W., "Chromatofocusing," in Protein Purification: Principles, High Resolution Methods, and Applications, J.–C. Janson and L. Ryden (eds.), VCH Publishers, New York (1989).

Karlsson, E., L. Ryden, and J. Brewer, "Ion exchange chromatography," in Protein Purification: Principles, High Resolution Methods, and Applications, J.–C. Janson and L. Ryden (eds.), VCH Publishers, New York (1989).

Lampson, G. P., and A. A. Tytell, "A simple method for estimating isoelectric points," *Analytical Biochemistry* 11:374–377 (1965).

Maskalick, D.G., and M.A. Abbott, "Choice of salt and flow rate can affect recovery of biomolecules in ion exchange chromatography," in Protein Purification: Micro to Macro, R. Burgess (ed.), Alan R. Liss, New York (1987).

Torres, A. R., B. E. Dunn, and S. C. Edberg, "Preparative HPLC of proteins on ion exchange columns with carboxymethyldextrans as displacers," *Journal of Chromatography* 316:125–132 (1984).

Torres, A. R., and E. A. Peterson, "Displacement chromatography of simple protein mixtures," *Journal of Biochemical and Biophysical Methods* 1:349–360 (1979).

Vermeulen, T., M. D. LeVan, N. K. Hiester, and G. Klein, "Adsorption and Ion Exchange," in Chemical Engineers' Handbook, 6th ed., R. H. Perry, D. W. Green, and J. A. Maloney (eds.), McGraw–Hill, New York (1984).

Yamamoto, S., K. Nakanishi, and R. Matsuno, Ion–Exchange Chromatography of Proteins, Marcel Dekker, New York (1988).

FURTHER READING

Regnier, F. E., and R. M. Chicz, "Ion exchange chromatography," in High Performance Liquid Chromatography of Biological Macromolecules: Methods and Applications, K. M. Gooding and F. E. Regnier (eds.), Marcel Dekker, New York (1989).

Tice, P. A., I. Mazsaroff, N. T. Liu, and F. E. Regnier, "Effects of large sample loads on column lifetime in preparative scale liquid chromatography," *Journal of Chromatography* 410:43–51 (1987).

CHAPTER 12

HYDROPHOBIC INTERACTION CHROMATOGRAPHY AND REVERSED PHASE CHROMATOGRAPHY

Hydrophobic interaction chromatography and reversed phase chromatography are separations based on the attraction between hydrophobic groups on the protein and a hydrophobic matrix. In hydrophobic interaction chromatography the sample is applied under conditions of high salt concentration and eluted under conditions of low salt concentration. In reversed phase chromatography the sample is applied in an aqueous buffer and eluted in an organic buffer. Various other compounds may be incorporated to improve the separation.

The distinction between the two methods is one of degree: The matrix in reversed phase chromatography is more highly hydrophobic than is that in hydrophobic interaction chromatography. In reversed phase chromatography the interaction between the protein and the matrix is usually too strong for the protein to be eluted with aqueous buffers. Organic solvents elute the protein by interacting with the protein or matrix or by disrupting the structure of the water. In hydrophobic interaction chromatography the interaction between the protein and the matrix is less strong, and aqueous buffers can be used for elution. In both systems the protein is distributed between the hydrocarbon–containing matrix and the polar mobile phase.

Hydrophobic interaction chromatography is applicable to most proteins although the degree of separation is often lower than a comparable ion exchange or affinity chromatography operation. To compensate, reversed phase operations may employ small particle sizes with high pressures.

Reversed phase is most effective with small proteins; large proteins tend to become denatured and may be difficult to elute, especially if the unfolding occurs within a pore of dimensions comparable to the protein. Proteins with a large number of hydrophobic residues may bind so tightly as to be impossible to elute. Antibodies, for example, tend to aggregate and become more difficult to handle following adsorption to a strongly hydrophobic matrix. Transmembrane proteins normally reside in an environment similar to the reversed phase matrix and may be particularly difficult to elute. The combination of high resolution, high capacity, and high yield frequently make reversed phase the method of choice for molecules of molecular weight less than 10,000 D. Hydrophobic interaction offers distinct advantages for unstable proteins. Where reversed phase requires a stable protein, hydrophobic interaction can be run in high concentrations of ammonium sulfate, a condition that favors the stability of many proteins, including glycoproteins from mammalian cells.

Hydrophobic interaction chromatography can also be used as a concentration technique, and if the sample contains a high salt concentration, such as the eluate from an ion exchange step, it can be applied directly to the column without first exchanging the buffer.

The term "reversed phase" comes from contrast with "normal phase" chromatography, a technique that uses a polar stationary phase and a nonpolar mobile phase, such as in the separation of small molecules by adsorption to silica, alumina, or glass in organic solvents such as chloroform, methylene chloride, or pyridine. Normal phase chromatography is a standard method in organic chemistry that has not found widespread use in protein purification.

12.1 MECHANISM

Proteins are made up of chains of amino acids, each with a side group attached, some of which are hydrophobic. The amino acids alanine, valine, leucine, and isoleucine have hydrocarbon chains attached; proline has a cyclic hydrocarbon attached; phenylalanine, tyrosine, and tryptophan have aromatic rings attached. Proteins order themselves in aqueous solution to achieve minimum free energy by burying most of their hydrophobic groups inside the protein and leaving the charged groups on the outside. Some hydrophobic groups will remain exposed on the outside, giving localized hydrophobic regions that are available for association with hydrophobic groups on the matrix.

The matrix may consist of a hydrophobic polymer such as styrene cross–linked with divinylbenzene, or it may be a hydrophilic polymer such as agarose with a hydrocarbon chain attached to provide hydrophobicity.

The exact mechanism by which proteins bind to hydrophobic matrices has been the subject of much debate. The empirical observation that the order of salts for both salting out (Hofmeister series; see Chapter 4) and hydrophobic interaction chromatography is approximately the same has led to a general belief that the mechanisms of these two processes are the same or at least similar in several aspects. Srinivasan and Ruckenstein (1980) have described the interactions as the result of the balance of van der Waals, electrostatic, and hydrophobic forces. The van der Waals interaction consists of three additive components. Constantly moving electrons create an instantaneous dipole in any molecule, which in turn generates an electric field that induces a dipole in neighboring molecules: Interaction between instantaneous dipoles results in London attraction, or dispersion; interaction between permanent dipoles is known as Keesom interaction, or orientation; interaction between a permanent dipole and an instantaneous dipole is known as Debye interaction, or induction. Orientation and induction are responsible for most of the van der Waals attraction between water molecules and are considered the major components of hydrogen bonding. Electrostatic interactions may be either repulsive or attractive (or both within a pair of protein molecules). Hydrophobic forces are repulsive over short ranges and include stearic repulsion, hydration repulsion, and Born repulsion. When the distance between two macromolecules decreases below the dimensions of the molecules, the number of conformations available to the molecules decreases, increasing the entropy and consequent repulsion between the molecules. Hydration repulsion results from the work required to remove the water of hydration surrounding a dipole (induced or permanent). Born repulsion occurs when the distance between surfaces is of atomic order, and it is due to the fact that electronic orbitals do not overlap.

The view of Srinivasan and Ruckenstein is somewhat at variance with the long–popular idea that the driving force is the increase in entropy that results from displacement of the water molecules away from the hydrophobic groups

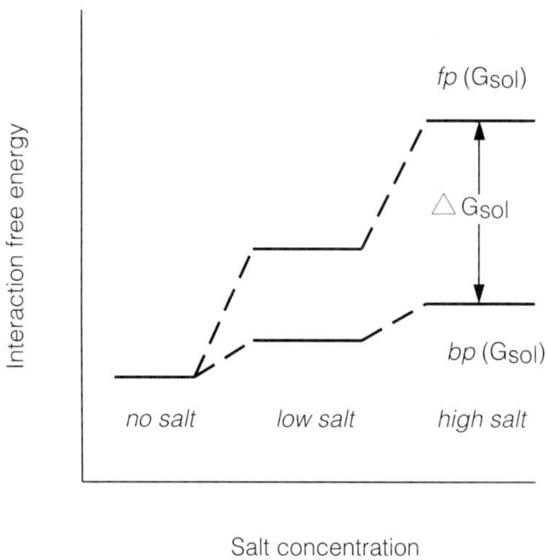

Figure 12.1. Diagram of relative change in free energy with change in salt concentration.

although those authors are in approximate agreement with Sinanoglu (1968), who ascribes the hydrophobic effect to the free energy change (the area of hydrophobic contact multiplied by the local surface tension of the fluid) that occurs upon association of hydrophobic groups. In either view the association between the protein and the solid phase is promoted by a decrease in the solubility of the protein in the fluid phase, which in effect causes the protein to precipitate onto the solid phase.

Arakawa (1986) has described the free–energy change associated with changing salt concentration. This is illustrated in Figure 12.1, where ΔG_{sol} is the change in free energy due to the interaction of the salts with the protein, $f_p(G_{sol})$ is the free energy resulting from interaction of the salts and the unbound (free) protein, and $b_p(G_{sol})$ is the free energy resulting from interaction of the salts and the bound protein. This diagram thus depicts the increase in the free–energy change with increasing salt concentration.

12.2 MATRICES

The binding of protein to a hydrophobic matrix is affected by the type of ligand, the ligand density in the matrix, the charge density of the ligand in the matrix, and the backbone material of the matrix.

Many different ligands are available, with a wide range of hydrophobic properties. Most ligands are hydrocarbon chains, some with phenyl, primary amine, or carboxyl groups attached. Table 12.1 lists some of the common ligands used in reversed phase and hydrophobic interaction chromatography.

Protein retention and column resolution vary directly with the density of the hydrophobic ligands and their hydrophobicity, a property that roughly correlates with the length of the ligand.

The major difference between hydrophobic interaction and reversed phase matrices is the relative hydrophobicity of the matrix, with reversed phase matrices having the greater hydrophobicity. Proteins tend to be held tightly

Table 12.1. Common ligands in hydrophobic interaction and reversed phase chromatography.

Ligand	Length, endgroup	Hydrophobicity, charge
C_n	n = 1, 2, 3, 4, 6, 8, 10, 18	Weak to strong
C_n-phenyl	Phenyl	Strong
C_n-amine	Primary amine	Basic (anion exchanger)

more by long hydrocarbon chains on the matrix than by short chains. And the more tightly the protein is held, the more drastic are the conditions needed to release it.

The retention of a protein is somewhat dependent on the ligand density and the size of the molecule. Fausnaugh et al. (1984) found that larger proteins were more strongly retained by matrices with higher ligand densities; small molecules showed less impact of ligand density on retention, a result that is probably due to the increased number of hydrophobic patches on large molecules. This correlation between size and binding affinity does not always hold. The ligand density also affects resolution, though not necessarily in a predictable way. Some proteins that bind tightly under reversed phase conditions have been found to bind only loosely under hydrophobic interaction conditions, and vice versa.

A number of hydrophobic ligands have an ionic group at the free end. This imparts ion exchange properties to the matrix, on top of the hydrophobic characteristics. The ionic charge density and the ligand density are both variables that influence the separation.

The matrices used for hydrophobic interaction and reversed phase chromatography include some that are also used for ion exchange chromatography as well as others that are distinctly hydrophobic. Table 12.2 lists several solid–phase matrices commercially available. The hydrophilic polymers find application in all sorts of chromatographic separations, depending on the functional group attached. The bonded silica consists of a silica base with a hydrophilic polymer coating to which the functional group

Table 12.2. Comparison of matrices for hydrophobic interaction and reversed phase chromatography.

Matrix	Ligands	Application
Hydrophilic:		
Agarose	C_1–C_{18}	hic, rpc
Bonded silica	C_3	hic
Ethylene glycol methacrylate	C_3, C_4	hic
Silica	C_1 to C_{18}	rpc
Hydrophobic:		
Styrene divinyl benzene	none	rpc
Teflon	none	rpc

is attached. Both Teflon and styrene divinylbenzene offer hydrophobic matrices without the use of hydrocarbon ligands. Silica is widely used in high–pressure systems, where the particle sizes are usually about 10 to 20 μm, with a range of 5 to 40 μm.

The silanol groups on the silica are coupled to the hydrocarbon, giving an ester linkage. Residual silanes that are not coupled to hydrocarbons may be capped to make them inert and unreactive. All other things being equal, capped and uncapped columns will give quite different performances, including a different order in which proteins are eluted.

Pore diameters of 25 to 30 nm (250 to 300 Å) are commonly used for large proteins although Kroeff et al. (1989) reported that a pore size of 12 to 15 nm (120 to 150 Å) was ideal for insulin, a small protein. The capacity of a matrix decreases with increasing pore size due to the lesser amount of resin within a particle of a given size. Nonporous particles less than 2 μm in diameter are also available for analytical columns and provide rapid separation but have low capacity. The absence of pores may make them attractive for very large proteins, which cannot diffuse well into small pores.

12.3 OPERATION AND APPLICATION

Operation of hydrophobic columns is very similar to that of ion exchange columns: The column is first equilibrated, the sample is adsorbed to the matrix, the product is eluted, and the column is regenerated to remove any remaining proteins and to prepare it for reuse. The column is then reequilibrated and another cycle started.

12.3.1 Hydrophobic Interaction

Proteins are eluted from a hydrophobic matrix by reduction of the strength of the hydrophobic bond through changes in the mobile phase or by alteration of the conformation of the protein. By regulation of the mobile–phase variables of salt concentration, salt type, and pH, the retention can be affected. The use of mild detergents or low concentrations of denaturants may disrupt the structure of the protein sufficiently to release the protein from the matrix.

At high ionic strengths the nature of the salt has considerable impact on the hydrophobic interaction between protein and matrix. The lyotropic series of Hofmeister (Chapter 4) ranks the salts as kosmotropes (water–structure forming) and chaotropes (water–structure breaking). Kosmotropes will favor adsorption of proteins to hydrophobic matrices; chaotropes will favor desorption.

Polarity–reducing compounds such as ethylene glycol have been found to greatly facilitate the elution of some proteins. Shaltiel (1984) described the use of 1 M urea and 0.4 M imidazolium citrate, pH 7.0, in the elution of glycogen phosphorylase–B. While care must be taken to avoid highly hydrophobic detergents such as sodium dodecyl sulfate (SDS), which will bind tightly to hydrophobic columns, non–ionic detergents such as CHAPS have been used successfully by Buckley and Wetlaufer (1989).

Different salts have different effects on the retention of proteins by hydrophobic interaction matrices. Fausnaugh et al. (1984) looked at the elution of proteins by a linear gradient decreasing from 1.0 M salt to zero salt and found that of the salts they tested the retention of protein increased in the order

$$NH_4SCN < NaCl < (NH_4)_2SO_4 < Na_2SO_4 < Na_3\text{citrate} \tag{12.1}$$

A more complete list is given in Chapter 4, with the salts ranked by order of molal surface–tension increments. It is apparent that the salt does more than simply exist as charged ions in solution, as suggested by classical electrostatics.

To the extent that hydrophobic interaction chromatography is a salting out, or precipitation, of the protein on the column, alterations in the solubility of the protein may affect its retention on a hydrophobic matrix although Fausnaugh et al. (1984) found no correlation between pH of buffer, isoelectric point, and retention for several proteins tested. (Even though buffer pH may not correlate well with isoelectric point, variation in the pH does affect the elution.) They also found that, the more hydrophobic the protein, the greater the resolution as measured by separation between successive peaks.

12.3.2 Reversed Phase

The retention of proteins on a reversed phase chromatography matrix is dependent on the relative hydrophobicity of the protein compared to the polarity of the fluid and solid phases. Protein adsorption on the solid phase is enhanced by such polar solutions as buffers and is decreased by such nonpolar organic solvents as acetonitrile. Elution occurs when the relative hydrophobicity of the protein is decreased by either decreasing the polarity of the fluid phase or increasing the polarity of the protein through the addition of an ion–pairing agent. Ion–pairing agents may also be used to enhance adsorption by decreasing the polarity of the protein. Some agents, such as trifluoroacetic acid, markedly improve the bandwidth and peak shape of proteins in reversed phase.

The retention of proteins on a reversed phase chromatography matrix can be controlled by changing the variables of the fluid phase, namely, solvent concentration, solvent type, and ion–pairing agent. In samples that have ionic properties the ion–pairing agent is especially important. Interactions between the sample and the solvent phase can be stronger than those between the sample and the nonpolar matrix.

Organic solvents can be classified by their ability to accept protons (basicity), donate protons (acidity), and relative dipole moment. Organic solvents can be ranked in order by the concentration at which they elute a protein as follows:

$$\text{tetrahydrofuran (THF)} < \text{acetonitrile} < \text{methanol} < \text{water} \tag{12.2}$$

That is, it takes a lower concentration of acetonitrile than of methanol to elute a protein. Table 12.3 lists several organic solvents used in reversed phase chromatography. A more complete list is given by Snyder and Kirkland (1979). Changes in relative retention are brought about more by changing the nature of the solvent (for example, from a proton acceptor, such as butanol, to a proton donor, such as water, or a solvent with a large dipole, such as acetonitrile) than by moving among homologs, for example, from propanol to ethanol.

Ion–pairing agents work to provide more retention and greater selectivity, particularly for highly acidic or basic samples. Snyder et al. (1988) have described two possible mechanisms for the action of ion–pairing agents:

> ...(a) the formation of an ion–pair between a sample ion and the ion–pairing agent in the mobile phase, followed by hydrophobic association of the ion–pair with the column stationary phase, or (b) a process whereby the ion–pairing agent first adsorbs to the surface of the stationary phase, and the sample ion is then retained by an ion exchange mechanism.

In the first case the ion–pairing agent combines with the protein to form a complex with increased hydrophobicity or decreased solubility in the fluid phase. In the second case the ion–pairing agent forms a complex with the hydrophobic matrix, changing the adsorption mechanism to ion exchange. Variation in the amount and type of ion–pairing agent shifts the controlling mechanism between ion exchange and hydrophobic interaction. Hexane sulfonate for basic samples and tetrabutylammonium (TBA) for acidic samples are examples of ion–pairing agents. Triethylamine (TEA) is sometimes added for improved resolution. Because of the ion exchange effects of ion–pairing agents, the resolution may be further influenced by alteration of pH and ionic strength. Many hydrophilic proteins chromatograph well in solutions of formic, acetic, or phosphoric acid at pH 2 to 4. Table 12.4 lists common ion–pairing agents.

Table 12.3. Organic solvents commonly used in reversed phase chromatography of proteins ranked in order of increasing polarity or chromatographic strength.*

Solvent	Polarity
Isooctane	0.1
Triethylamine	1.9
Butanol	3.9
Isopropanol	3.9
n-Propanol	4.0
Tetrahydrofuran (THF)	4.0
Ethanol	4.3
Methanol	5.1
Acetonitrile	5.8
Water	10.2

* From Snyder (1978).

Changes in the buffer that affect the solubility of the protein will affect its binding to a hydrophobic matrix. The addition of salt to the buffer will increase or decrease the affinity of the protein for the matrix, depending on whether the salt decreases or increases the solubility of the protein in solution. Similarly, increasing the ionization of the protein (by moving the pH away from its isoelectric point) may increase the protein solubility and decrease its relative affinity and adsorption on the matrix.

Table 12.4. Ion pairing agents commonly used in reversed phase chromatography, listed from top to bottom in order of increasing hydrophobicity.

Ion-pairing agent	Hydrophobicity
Anionic:	
Phosphoric acid*	Zero
Formic acid	Low
Acetic acid	
Trifluoroacetic acid (TFA)	
Heptafluorobutyric acid (HFBA)	
Hexane sulfonate	
Dodecyl sulfonate	High
Cationic:	
Ethylenediamine	Low
Triethylamine (TEA)	
Triethylammonium phosphate	
Tetrabutylammonium (TBA)	High

* Though commonly listed as an ion pairing agent, phosphoric acid has zero hydrophobicity and is an ion suppressing agent.

Wehr et al. (1989) have found that the addition of small amounts of detergent served to improve resolution, with the greatest effect derived from zwitterionic detergents on large and hydrophobic proteins and little effect derived from ionic detergents. The resolution of proteins that aggregated at low detergent concentrations was improved with increased detergent concentration.

It is important to keep in mind that, when we work with proteins, the complexity of the environment provides many exceptions to the general rules. Binding between protein and matrix occurs at a locally small region of the molecule, and general properties of the protein, such as isoelectric point, may not reflect the separation that can be obtained under any specific circumstances.

12.4 OPTIMIZATION AND SCALE UP

The selection of a hydrophobic or reversed phase matrix can be made in a manner similar to that described previously for ion exchange. A sample can be applied to multiple small columns containing a variety of matrices of varying hydrophobicity, which are then washed with the equilibration buffer. Analysis of this flow–through material will indicate the proteins that bind to each type of resin. Once the proper resin has been selected, the elution buffer composition can be varied to provide the best resolution.

The matrix of choice is not necessarily the one with the shortest hydrocarbon chain. When more than one matrix is found to bind the protein, the matrix that offers the greatest separation between the product and contaminants will be the preferred matrix. Multiple matrices and elution schemes provide a number of alternatives for optimization of the resolution of a hydrophobic or reversed phase step.

Fausnaugh et al. (1984) recommend that initial experiments be performed with a salt of intermediate salting–out ability at a concentration of at least 1 M. The retention can be increased by increase of the salt concentration or by a salt of higher salting–out ability; the retention can be decreased by decrease of the salt concentration or by a salt of lower salting–out ability.

One of the trade–offs that must be considered during optimization is between recovery and resolution. The nature of hydrophobic matrices is such that they bind some proteins very tightly. Selection of an ion–pairing agent should be made with consideration for recovery as well as resolution. The recovery of proteins from a hydrophobic matrix tends to increase with temperature up to about 60°C.

The solvents that are used in reversed phase chromatography are flammable and operation at large scales requires special equipment to prevent explosion or fire. In addition, many solvents and ion–pairing agents are toxic, and proper ventilation must be provided. Following chromatography, the solvent must be removed from the product and recycled; these steps may influence the selection of the solvent or the ion–pairing agent. The gain in yield that results from the use of a particular solvent may be totally lost in the diafiltration that is required to prepare the product for the next step in the purification.

REFERENCES

Arakawa, T., "Thermodynamic analysis of the effect of concentrated salts on protein interaction with hydrophobic and polysaccharide columns," *Archives of Biochemistry and Biophysics* 248:101–105 (1986).

Buckley, J. J., and D. B. Wetlaufer, "Use of the surfactant 3–(3–cholamidopropyl)–dimethyl–ammoniopropane sulfonate (CHAPS) in hydrophobic interaction chromatography of proteins," *Journal of Chromatography* 464:61–71 (1989).

Fausnaugh, J. L., L. A. Kennedy, and F. E. Regnier, "Comparison of hydrophobic interaction and reversed phase chromatography of proteins," *Journal of Chromatography* 317:141–155 (1984).

Kroeff, E. P., R. A. Owens, E. L. Campbell, R. D. Johnson, and H. I. Marks, "Production scale purification of biosynthetic human insulin by reversed phase high performance liquid chromatography," *Journal of Chromatography* 461:45–61 (1989).

Shaltiel, S., "Hydrophobic chromatography,", *Methods in Enzymology* 104:69–96 (1984).

Sinanoglu, O., in Molecular Associations in Biology, B. Pullman (ed.), Academic Press, New York (1968).

Snyder, L. R., J. L. Glajch, and J. J. Kirkland, Practical HPLC Method Development, John Wiley & Sons, New York (1988).

Snyder, L. R., and J. J. Kirkland, Introduction to Modern Liquid Chromatography, 2nd ed., John Wiley & Sons, New York (1979).

Snyder, L. R., "Classification of the solvent properties of common liquids," *Journal of Chromatographic Science* 16:223–234 (1978).

Srinivasan R., and E. Ruckenstein, "Role of physical forces in hydrophobic interaction chromatography," *Separation and Purification Methods* 9(2):267–370 (1980).

Wehr, C. T., R. P. Lundgard, and K. D. Nugent, "Hydrophobic proteins: a challenge for reversed phase HPLC," *LC–GC* 7(1):32–37 (1989).

FURTHER READING

Chang, S. H., K. M. Gooding, and F. E. Regnier, "Use of oxiranes in preparation of bonded phase supports," *Journal of Chromatography* 120:321–333 (1976).

Hearn, M. T. W. (ed.), Ion–Pair Chromatography, Marcel Dekker, New York (1985).

Hjerten, S., "Hydrophobic interaction chromatography," *Advances in Chromatography* 19:111–123 (1981).

Melander, W., and C. Horvath, "Salt effects on hydrophobic interactions in precipitation and chromatography of proteins: an interpretation of the lyotropic series," *Archives of Biochemistry and Biophysics* 183:200–215 (1977).

Regnier, F. E., and K. M. Gooding, "High performance liquid chromatography of proteins," *Analytical Biochemistry* 103:1–25 (1980).

Rivier, J., R. McClintock, R. Galyean, and H. Anderson, "Reversed phase high performance liquid chromatography: preparative purification of synthetic peptides," *Journal of Chromatography* 288:303–328 (1984).

CHAPTER 13

AFFINITY CHROMATOGRAPHY

Affinity chromatography relies on a specific interaction between the product protein and the solid phase to effect a separation from contaminants in the feed. This specific reaction is what makes affinity chromatography such a strong tool for purification. Rather than relying on relatively small differences between the product and contaminants that lead to purifications of only fewfold improvement, affinity reactions allow hundred– to thousandfold purifications within a single step. The rapid purification—compared to less specific methods—accounts for a large measure of the popularity of affinity chromatography. Affinity chromatography, like ion exchange, is a concentrating technique, handling large volumes of dilute media and delivering a small volume of concentrated product.

The literature on affinity chromatography fills a modest–size library. Wilchek et al. (1984) has reviewed almost five hundred applications in protein purification. Rather than trying to cover everything here, this chapter will concentrate on the application of affinity chromatography to large–scale operations, including techniques for scale up.

The term "affinity" has been used in previous chapters to describe the interaction between proteins and the solid phase; affinity as applied in the phrase "affinity chromatography" also refers to the interaction between protein and matrix. However, affinity chromatography applies particularly to strong and highly specific interactions.

13.1 MECHANISM

Affinity chromatography consists of the same steps as ion exchange or hydrophobic interaction chromatography, namely, adsorption, elution, and regeneration. These steps are illustrated in Figure 13.1. The differences among these types of chromatography lie in the nature of the functional group (called the "ligand" in affinity chromatography). Adsorption occurs when the product molecule binds to the ligand. Impurity molecules are washed from the resin, and the product is eluted from the column. In some cases the affinity column does not need regeneration.

As outlined by Absolom (1981), for an affinity chromatography step to be successful, two conditions must be met: A specific ligand that binds the product protein must be available in a form attached to a solid phase, and a means must be available for desorbing the protein from the ligand.

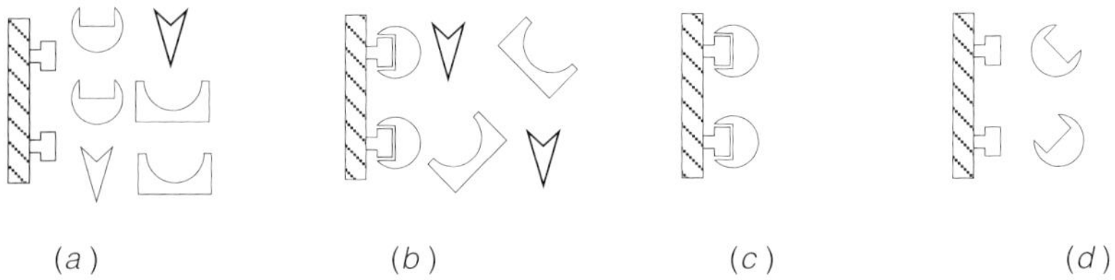

Figure 13.1. Schematic diagram of steps in affinity chromatography: (*a*) addition of sample, (*b*) adsorption of product to ligand, (*c*) removal of nonadsorbing contaminants, (*d*) elution of product and regeneration of matrix.

13.1.1 Affinity Ligands

Ligands vary in their specificity for proteins, and both highly specific and less–specific ligands are used in affinity chromatography. Highly specific ligands will bind only one type of protein, while general ligands will bind a whole class of proteins (for example, antibodies).

General ligands can be divided, as in Table 13.1, by the class of protein they bind. General ligands bind to a structure that multiple proteins have in common. The anthraquinone dye Cibacron blue F3G–A acts much like ATP, NAD, or NADP and thus can bind a wide range of proteins.

The general class of thiol ligands represents a special type of affinity chromatography, in which covalent bonds are formed between the ligand and protein. Belew et al. (1987) have attached mercaptoethanol to Sepharose to produce a general adsorbent for antibodies.

The binding or holding of a metal ion between two molecules is called "chelation," and forms the basis of a useful separation. The matrix is prepared by reaction with diacetic acid (or some other metal–chelating group), and metal ions are chelated to the column, as described by Porath et al. (1975). Proteins that chelate with such heavy metals as copper, zinc,

Table 13.1. Various ligands and the general classes of proteins they bind.

Ligand	Application
General ligands:	
Anthraquinone dyes Cibacron blue Procion red	Kinases, dehydrogenases
Heparin	Lipoproteins
Lectins Concanavalin A Lentil lectin	Glycoproteins, polysaccharides
Metal chelates	Histidine, cysteine groups
Protein A Protein G	Antibodies
Thiol	Sulfhydryl groups
Specific ligands:	
Enzymes	Substrates, inhibitors
Receptors	Hormones
Antibodies	Antigens

cadmium, nickel, and cobalt, through the imidazole and thiol groups of histidine and cysteine, will adsorb to the matrix. With transition metals selective adsorption of histidine and cysteine is favored at pH 6 to 8, while at higher pH less specific adsorption of amino groups takes place. Elution can be effected by a reduction in the pH, an increase in ionic strength, or displacement with EDTA (around 50 mM). Metal chelate chromatography has become an important and fast–growing method for protein purification.

The absolute value of the affinity between the product protein and the ligand is important when we choose a ligand for affinity chromatography. Based on the equilibrium between adsorbed and desorbed protein, the affinity can be quantified in terms of an adsorption equilibrium constant having units of liters per mole. Absolom (1981) recommends an affinity constant between 10^4 and 10^8 l/mole. At the high end of the range the elution of the product will require harsh conditions; at the low end of the range binding may not be strong enough to adsorb all the protein in solution.

The selection of the proper ligand may require trial and error. Antibodies used as ligands for antigens vary widely in relative affinities. One may bind so strongly that elution cannot be accomplished without denaturation of the product. Another may bind so loosely as to be eluted in the wash step.

Spacers are often necessary to facilitate binding. The spacer serves to move the ligand off the surface of the solid support, making it more accessible to the protein and overcoming the stearic hindrance between the product and the ligand caused by the matrix. Hydrocarbon chains of from five to ten carbons are frequently used: for example, diaminohexane and aminohexanoic acid. Care must be taken to ensure that the spacer does not contribute to nonspecific binding through ionic or hydrophobic interactions.

13.1.2 Coupling Chemistries

Successful affinity chromatography requires that the ligand remain active after it is coupled to the matrix. Care must be taken at each of the reaction steps to protect the ligand activity. The ligand is bound to the matrix in two steps: The matrix is first activated; then the ligand reacts with the activated matrix. If a spacer is used, it may be attached to the matrix and the ligand attached to it, or the spacer may be attached first to the ligand and the ligand–spacer complex attached to the matrix.

Coupling is accomplished by introduction of an electrophilic group onto the matrix for reaction with the nucleophiles available on the ligand. Active groups available for coupling reactions on both the solid phase and on the ligand include hydroxyl groups, carboxyl groups, amino groups, and sulfhydryl groups.

Many coupling chemistries are available; the suitability of each depends on the requirements of the particular application. Some commonly used methods are listed in Table 13.2 and include those involving cyanogen bromide, carbodiimide, tresyl (sulfonyl) chloride, and oxirane. The bonds with greatest stability are the secondary amines, followed in descending order of stability by urethane and isourea bonds. The more stable the bond, the less leaching, or loss of ligand, will occur. Such loss leads to reduced capacity and also contaminates the product.

Cyanogen bromide reacts with hydroxyl groups on the matrix to form a reactive group that will bind to amines through the formation of a isourea bond. This is the most widely used method of coupling, more from a historical reason than anything else. The stability of this bond as measured by ligand leaching is poor. Scouten (1987) suggests that use of triethylammonium nitrile improves the efficiency of the activation.

Sulfonyl chlorides, such as tresyl (trifluoroethylsulfonyl) chloride and tosyl (toluene sulfonyl) chloride, allow efficient coupling near neutral pH and a temperature of 4°C. The buffer should not contain nucleophiles (free amino groups) that could compete with the ligand for reaction sites on the matrix.

Table 13.2. Coupling chemistries used for binding of ligands to matrices for affinity chromatography.

Method	Matrix group	Ligand group	Bond type
Bisoxirane	Hydroxyl	Amine, thiol	Secondary amine
Carbodiimide	Carboxyl	Amine	Amide
Carbonyl diimidazole	Hydroxyl	Amine	Urethane
Cyanogen bromide	Hydroxyl	Amine	Isourea
Divinylsulfone	Hydroxyl	Amine	Secondary amine
2-Fluoro-3-methyl pyridinium tosylate	Hydroxyl	Amine, thiol	Secondary amine
N-hydroxysuccinimide	Hydroxyl	Amine	Urethane
Tresyl (sulfonyl) chlorides	Hydroxyl	Amine, thiol	Secondary amine

The N–hydroxysuccinimide and divinylsulfone coupling methods provide spacer arms between the ligand and matrix, while the other methods shown in Table 13.2 do not. Several of these methods have been reviewed by Scouten (1987).

13.2 MATRICES

The solid support is an important part of the affinity chromatography operation and should incorporate the following characteristics: It should not react in a nonspecific fashion with *any* proteins in the sample; it should have a strong enough structure to provide uniform and rapid flow conditions; it should be resistant to chemical or microbial degradation; it should have a sufficiently large pore size so that the protein can reach the ligand; it should have a high capacity for binding.

Nonspecific reactions include ionic interactions between charged groups, hydrophobic interactions, hydrogen bonding, and other types of general association between proteins and the matrix. Many hydrophilic matrices that minimize nonspecific binding are organic polymers having little structural strength. A soft matrix will deform under the pressure of large columns or high flow rates and be unsuitable for large–scale use. The more highly cross–linked resins show better mechanical stability. Polysaccharides may be susceptible to microbial attack, with degradation of the structure as a result. The matrix and ligand must be resistant both to antibacterial chemicals that may be necessary for the storage of columns and to all process buffers that may be applied. Proteins are large molecules and require large pores in the matrix for free diffusion to and from the ligands. A small pore size may allow binding of large amounts of a small ligand yet show low capacity for product because of exclusion of the protein from the matrix. This requirement competes with the requirement for structural integrity: Large pore gels tend to have softer structures. The capacity of the matrix depends on several factors, including the number of functional groups available for binding and the availability of ligands for adsorption of product.

Several of the support matrices that are available commercially are listed in Table 13.3. Agarose has been in use for many years and is the most widely used matrix. It is available with many ligands attached and has good flow properties. Agarose is a copolymer of alternating D–galactose and 3–6–anhydro–L–galactose subunits joined at the 1–3 and 1–4 positions. Agarose may be highly cross linked for increased mechanical stability, although with some loss of ligand binding capacity. The large pore structure obtainable with agarose allows its use with very large molecules. Cellulose is available in beaded form, but its many functional groups lead to nonspecific binding, especially if oxidative coupling methods are used since these lead to formation of aldehyde and carboxyl groups on the cellulose matrix. Cellulose is not suitable for low–pH elution schemes, as the glycosidic bond is susceptible to acid hydrolysis. Dextran is cross linked through alpha–1,6–glucosidic linkages to form an insoluble matrix. Polyacrylamide is stable over a wide pH range (pH 1 to 10); it is inert biologically, resistant to chemical and microbial attack, and does not contain ionic groups. If the gel is only lightly cross linked, derivitization may result in marked shrinkage of the matrix. Mosbach et al. (1984) describe "tailoring" of acrylamide supports for particular applications by varying their chemical stability and degree of porosity. Fluorocarbons are extremely inert

Table 13.3. Representative matrices for affinity chromatography.

	Matrix	Brand name	Source
Polysaccharides:			
	Agarose	Sepharose	Pharmacia
	Cellulose	Cellulofine	Amicon
	Dextran	Sephadex	Pharmacia
Synthetic polymers:			
	Acrylimide		
	Fluorocarbons	Perflex	DuPont
	Vinyl	Toyopearl HW	TosoHaas
Inorganic materials:			
	Silica	HiPAC	ChromatoChem

chemically and give good structural support and flow characteristics. Polyvinyl also has good flow and structural properties. Silica has been used as a support for affinity chromatography at high pressures. Ernst–Cabrera and Wilchek (1987) have reviewed the application of affinity chromatography in HPLC.

13.3 OPERATION AND APPLICATION

To meet the goal of an affinity chromatography step (to maximize separation while minimizing cost), each step of the operation must be optimized. The factors that affect adsorption are similar to those that affect desorption: pH, ionic strength, temperature, the presence of competing species, and flow characteristics. Between adsorption and elution of the product, extraneous contaminants that are retained by nonspecific adsorption (or in some cases less–specific adsorption) are usually washed from the column, either with the same buffer used during adsorption or with modified conditions such as altered pH or ionic strength.

Elution is a critical step in the success of an affinity operation. For most products not only must the protein be recovered with high yields, but it must also be biologically active. The development of an elution scheme for affinity chromatography must be done empirically. The better the protein–ligand bond is understood, the better one can predict what conditions are required.

Elution methods may be either general or specific. General methods include changes in temperature, ionic strength, pH, and solvent composition and the use of chaotropic agents. Specific methods include competitive molecules, such as a soluble ligand to compete for the immobilized ligand or a soluble counterligand to displace the product. The forces binding the product to the ligand that must be disrupted include electrostatic and hydrophobic interactions and hydrogen bonding.

Elution may be effected by changing the pH, which alters the extent of ionization of the protein. A solution of 0.1 M acetic acid or 0.1 M ammonium hydroxide may be sufficient, or stronger conditions, such as pH 1.5 may be needed. The stabilities of the product and ligand will limit the pH range that can be used.

The high resolution that makes affinity chromatography so desirable may be achieved through selective adsorption and general desorption, such as a change in the ionic strength of the buffer—NaCl at 1 M or less, for example. Less–harsh elution conditions may aid the preservation of biological activity. However, high association constants ($K > 10^6$) may require a chaotropic agent such as 6 M guanidine, 3 M sodium thiocyanide, or 3 M sodium iodide. Urea (8 M) can also serve as an elution agent. These agents are usually used at neutral pH.

The addition of organics, such as ethanol (5 to 10 percent), dioxane (10 percent), or ethylene glycol (25 to 50 percent), to the buffer will reduce hydrophobic interactions and may facilitate elution.

A combination of a disrupting agent and a change in the pH or ionic strength may be necessary to elute strongly bound moieties. In the case of a metal–chelate column, a change in pH or ionic strength or displacement with EDTA or a metal ion will elute the product.

Specific displacement molecules may work for many different types of product. Free haptens or analogs may serve to displace antigens from an antibody column. Substrates may displace enzymes. Kopperschlager et al. (1982) recommend ATP or NAD or analogs for the elution of kinases and dehydrogenases from anthraquinone dyes such as Cibacron blue.

During the development of an elution scheme, Wilchek et al. (1984) recommend trial of general means first, such as minor changes in ionic strength and pH to remove any contaminants attached by nonspecific binding, followed by more specific eluants to displace the product. Absolom (1981) lists potential eluants for several classes of general ligands.

13.4 OPTIMIZATION AND SCALE UP

The optimal aspect ratio, or ratio of column height to width, will depend on the kinetics of the adsorption reaction. If the adsorption proceeds rapidly to completion, a shorter contact time is sufficient for complete binding. A slower reaction will need a longer time of contact between the protein and ligand for complete adsorption. The residence time is the ratio of the column length and the linear speed of the fluid phase (flow rate divided by cross–sectional area). Operations with high adsorption rates can use a high flow rate and short column. Low adsorption rates will call for a lower flow rate and longer column. Linear flow rates are typically of the order of 1.2 to 12 cm/h. A high association equilibrium constant will dictate a smaller aspect ratio.

The cost of an affinity step is usually controlled by the cost of the ligand and associated matrix, which may be very expensive if the column cannot be reused. Some columns can be used many times, leading to a significant reduction in the volume of resin needed. Affinity chromatography, because of the high cost of the solid phase, is especially suited to automation, with a small column run many times through the same cycle. The ability to reuse the column depends on how well bound substances can be removed and how well the ligand activity can be preserved.

13.5 OTHER AFFINITY METHODS

Ligands may also be attached to membranes. Most frequently used are isotropic microfiltration membranes. These offer the advantage of reduced mass–transfer resistance because, instead of the molecule's diffusing into the bead, diffusion is restricted to the boundary layer surrounding the ligand. Flow is usually through the membrane, perpendicular to its surface, an arrangement that brings the sample into contact with the ligand very quickly. Ligands may be attached throughout the body of the membrane and not just on the surface.

REFERENCES

Absolom, D. R., "Affinity chromatography," *Separation and Purification Methods* 10(2):239–286 (1981).

Belew, M., N. Juntti, A. Larsson, and J. Porath, "A one–step purification method for monoclonal antibodies on salt–promoted adsorption chromatography on a thiophilic adsorbent," *Journal of Immunological Methods* 102:173–182 (1987).

Ernst–Cabrera, K., and M. Wilchek, "From affinity chromatography to HPAC," in Protein Purification: Micro to Macro, R. Burgess (ed.), Alan R.Liss, New York (1987).

Kopperschlager, G., H. J. Bohme, and E. Hofmann, "Cibacron Blue F3G–A and related dyes as ligands in affinity chromatography," *Advances in Biochemical Engineering* 25:101–145 (1982).

Mosbach, R., A. C. Koch–Schmidt, and K. Mosbach, "Immobilization of enzymes to various acrylic copolymers," *Methods in Enzymology* 104:53–65 (1984).

Porath, J., J. Carlsson, I. Olsson, and G. Belfrage, "Metal chelate affinity chromatography, a new approach to protein fractionation," *Nature* 258:598–599 (1975).

Scouten, W. H., "A survey of enzyme coupling techniques," *Methods in Enzymology* 135:30–65 (1987).

Wilchek, M., T. Miron, and J. Kohn, "Affinity chromatography," *Methods in Enzymology* 104:3–55 (1984).

FURTHER READING

Graves, D. J., and Y. T. Wu, "The rational design of affinity chromatography separation processes," *Advances in Biochemical Engineering* 12:219–252 (1979).

Kagedal, L., "Immobilized metal ion affinity chromatography," in Protein Purification: Principles, High Resolution Methods, and Applications, J.–C. Janson and L. Ryden (eds.), VCH Publishers, New York (1989).

Ryden, L., and J. Carlsson, "Covalent chromatography," in Protein Purification: Principles, High Resolution Methods, and Applications, J.–C. Janson and L. Ryden (eds.), VCH Publishers, New York (1989).

Scouten, W. H., Affinity Chromatography, John Wiley & Sons, New York (1981).

Tarnowski, S. J., S. K. Roy, R. A. Liptak, D. K. Lee, and R. Y. Ning, "Large–scale purification of recombinant human leukocyte interferons," *Methods in Enzymology* 119:153–165 (1986).

Yang, C. and G. T. Tsao, "Affinity Chromatography," *Advances in Biochemical Engineering* 25:19–42 (1982).

CHAPTER 14

GEL FILTRATION CHROMATOGRAPHY

Gel filtration chromatography is a process for the separation of molecules on the basis of size. Known also as "gel permeation chromatography" and "size exclusion chromatography," its ease of use and distinct basis for separation have made it a commonly preferred technique.

Unlike most chromatographic methods, gel filtration chromatography does not rely upon the adsorption of protein to a solid phase. In gel filtration chromatography, just as in adsorption chromatography, the various components of the sample are differentiated by the time they spend with the solid phase; but that time is not determined by the formation of weakly interacting bonds between the protein and the matrix. Rather it is determined by the extent to which the molecules diffuse in a random pattern into and out of the pores of the solid matrix. Those molecules which are small enough to diffuse an appreciable distance into the matrix will be retarded in their passage through the chromatography column; thus the retardation results in a separation in space and time between large and small molecules.

There are several applications for gel filtration chromatography, including, buffer exchange, desalting, and separation of product from contaminants of different size. Gel filtration chromatography may also be used as an analytical method to determine the molecular weight of a particle or a molecule (see Chapter 4) or to determine the molecular weight distribution of a series of synthesized molecules. Gel filtration chromatography may also be used to separate cells and cellular debris from soluble proteins.

Gel filtration chromatography is very useful in the desalting of protein solutions. In this case, a small pore size medium is preferred, and a short column is all that is required. The sample is applied to the column and eluted with the desired buffer. The salt ions in the sample are retained in the gel filtration (desalting) medium while the protein is recovered in the new buffer.

Gel filtration is a convenient way to remove DNA from protein solutions, as DNA tends to polymerize in low–salt buffers, passing through a gel filtration column as a large–molecular–weight aggregate.

Gel filtration chromatography columns can be run in the presence of harsh reagents, including urea, guanidine hydrochloride, and detergents. Gel filtration provides a means for separation of products that, because of insolubility, may not be amenable to other chromatographic techniques, such as ion exchange or affinity chromatography.

14.1 MECHANISM

Gel filtration chromatography is a process in which large molecules traverse the length of a column faster than small molecules do. Separation in gel filtration is a function of the ability of the various molecules in the sample to enter the pores of the stationary phase. Figure 14.1 shows schematically the exclusion feature of gel filtration chromatography media.

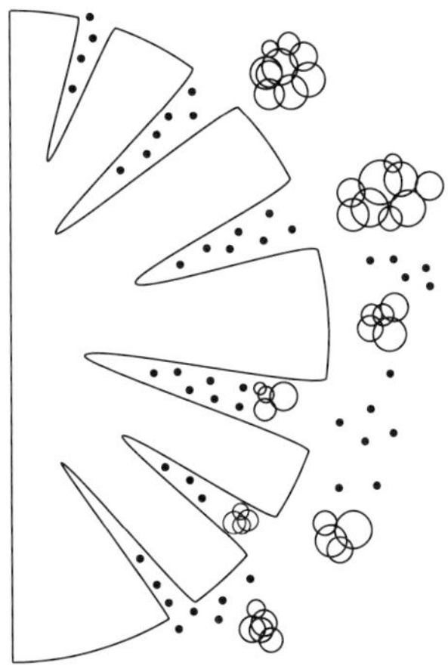

Figure 14.1. Schematic diagram, showing the extent of exclusion by size of molecules from the pores of the matrix.

The length of time a molecule takes to migrate through the column depends on its size. Large molecules are excluded entirely and move with the fluid front; intermediate ones enter part way into the pores and are retarded to a greater or lesser extent; small ones take considerably longer to pass through the column. As a molecule moves down a column, it diffuses into the pores and then back out. This diffusion, which is random in direction, takes time.

The farther the molecule diffuses into the pores, the longer it takes to move down the column. Therefore the smaller molecules, which diffuse farther into the matrix, take longer to migrate through the column.

Implicit in the discussion in this chapter is the assumption that the size of a molecule is proportional to its molecular weight. For globular proteins this is basically true. For linear or denatured proteins or proteins associated with detergent molecules this may not be so, and such molecules may, because of their ability or inability to enter narrow pores, chromatograph as though they were much smaller or larger than when they are in their native state.

A gel filtration step may be characterized in several ways: One commonly used method is in terms of the elapsed time between application of the sample and the elution of each species from the column, that is, the retention time. For constant flow rates, the retention time is directly proportional to the elution volume.

For ease of discussion, the volume within a gel filtration column may be divided into four classifications, as shown in Figure 14.2: The first is the actual volume taken up by the solid phase; this is the volume into which no protein can migrate. The second classification is the void volume; this is the general large channel area between medium particles. The third volume classification is the film thickness of solution surrounding the particles; this is the film layer through which molecules must diffuse to get to the pores of the particles. The fourth volume classification is the pore volume, the actual volume of liquid within the pores of the particles.

Equation 10.10 has been rearranged by Fisher (1980) to give

$$K = \frac{V_e - V_v}{V_s} \tag{14.1}$$

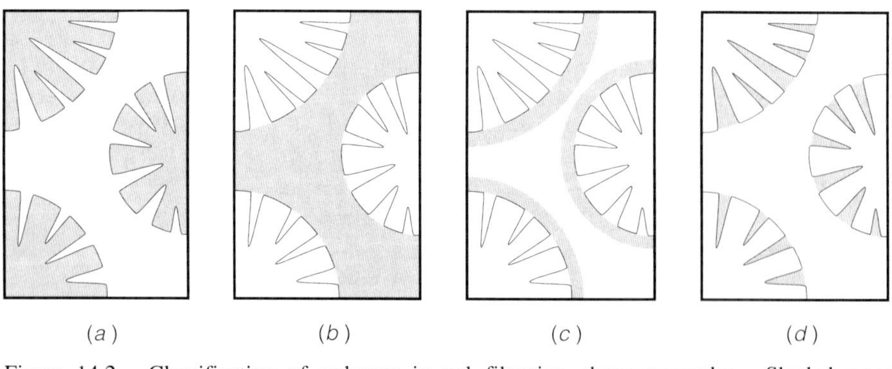

(a) (b) (c) (d)

Figure 14.2. Classification of volumes in gel filtration chromatography. Shaded areas represent (*a*) solid-matrix volume, (*b*) void volume, (*c*) film volume, (*d*) pore volume.

where V_e is the elution volume, V_v is the void volume, and V_s is the volume of the stationary phase. This definition of the distribution coefficient represents the ratio of the pore volume to the stationary–phase volume and thus represents the fraction of the stationary phase into which a given solute species may diffuse. Because it is difficult in practice to measure the volume of the stationary phase, the total volume minus the void volume, $V_t - V_v$, is often substituted for V_s. (The term $V_t - V_v$ includes the volume of the solid–phase matrix and therefore is not truly accessible to solute molecules). The distribution coefficient is useful in comparing the elution of different species or of the same species in different media.

The performance of a column may be described in terms of its ability to separate standard substances or in terms of an empirical measurement of the number of "theoretical plates." The latter is extremely useful for comparing and monitoring the quality of the column packing, and it is described in detail later in the section on scale up.

Several factors influence the quality of the separation in gel filtration chromatography. These include the medium (that is, pore size distribution and uniformity of packing), the velocity of the mobile phase, and the sample load, including both volume and mass concentration.

14.2 MATRICES

Several factors influence the choice of matrix, including pore size distribution, structural properties, and interactions (other than diffusion) between the sample and the matrix.

Media are characterized by the pore size distribution, or the number of each size of pore. The fractionation range of a matrix is the range of sizes of molecules that can diffuse, to a greater or lesser degree, into the pores of the matrix to provide a useful separation. Molecules that are too large to enter the pores are excluded from the matrix and will elute after a volume equivalent to the void volume has passed through the column. Small molecules, such as salt ions, will enter the matrix freely, resulting in severe retardation of elution time. The useful fractionation range is the range of molecule sizes that are retarded somewhat by diffusion into the matrix but not delayed as much as the salt ions. The exclusion limit is the size of molecule that is just too large to enter the matrix pores.

In the selection of a gel filtration medium the pore size of the medium has the largest effect on the resolution of the separation. In general, gel filtration operations are most successful when the product is included in the matrix in the middle of the fractionation range of the matrix. When a very large molecule is to be purified, it may occasionally be desirable to exclude the product from gel filtration medium pores entirely and allow the product to travel through in the void–volume fraction. Only in the case where there are no other particles larger than the product will this be an efficient strategy.

For media with equivalent porosities, that is, with similar numbers of pores per unit volume of medium, a narrower pore size distribution will better resolve proteins of near molecular weights. If the molecular weights of the contaminants are widely divergent from that of the product, then a wider pore size distribution may be preferred. A narrow pore size distribution gives higher selectivity, while a broad pore size distribution gives increased separation range.

The properties of the matrix material will determine the nature of the pores as well as the mechanical strength. Table 14.1 lists some commercial matrices and their suppliers. Other materials are reviewed by Hagel (1988).

Many gel filtration matrices are deformable under pressure, a condition that limits not only the maximum flow rate but also the maximum column height. Multiple columns are frequently run in series to obtain the desired resolution without crushing the bottommost gel. The greater the cross linking, the firmer the gel and the less deformation under pressure.

In the ideal situation the only effects that contribute to performance in gel filtration columns are steric effects; that is, the size of the molecule strictly determines the position at which it elutes. However, with nonideal systems, which is the case in all real–life situations, other factors come into play: These include interactions between the matrix and the sample species as well

Table 14.1. Gel filtration media.

Matrix	Brand name	Supplier
Acrylamide	Bio-Gel P	BioRad
Agarose	Sepharose	Pharmacia
	Superose	Pharmacia
	Ultrogel	IBF Biotechnics
	Bio-Gel A	BioRad
Dextran	Sephacryl	Pharmacia
	Sephadex	Pharmacia
Polyvinyl	Toyopearl HW	TosoHaas
Silica	Zorbax	DuPont

as interactions among species in the sample. Interactions between the protein molecules chromatographed and the solid–phase matrix are deemed undesirable in gel filtration chromatography because they are unpredictable. A single column that separates on more than one basis may actually provide improved resolution; such combinations are usually found by serendipity and are difficult to implement intentionally.

Some media for gel filtration chromatography exhibit sorptive characteristics under certain conditions. These may be weakly electrostatic, leading to ion exchange responses, or weakly hydrophobic, leading to hydrophobic interactions. Weak electrostatic interactions can usually be avoided by the addition of salt in the elution buffer. Hydrophobic nteractions are avoided by the use of less salt in the elution buffer or by the inclusion of a solvent such as ethanol.

The molecular–weight exclusion limits for gel filtration matrices are typically measured by polyethylene glycol and dextran. Because proteins are not rigid spheres, an exclusion limit is only an approximation to the molecular weight of a protein that would be prohibited from entering the pore of the matrix. Gel filtration columns may be calibrated by molecules with known molecular weight, such as electrophoresis standards. The use of materials such as blue dextran to calibrate columns is not recommended because of the tendency of such dyes to bind to some columns and subsequently leach out into the product.

14.3 OPERATION AND APPLICATION

The nature of the sample will affect the resolution of a gel filtration operation. Bulk characteristics, such as viscosity, concentration, and volume, play a role as well as solubility and molecular interactions.

The partition isotherms in gel filtration are usually quite linear, and therefore to a large extent gel filtration is independent of the sample concentration. However, high–viscosity samples will lead to irregular flow patterns; as the high–viscosity material is diluted, any variations in the thickness of the sample application across the surface of the column will be magnified. If the dilution process requires energy to break an association bond between molecules, the degree of resolution will be decreased. Anonymous (1982) recommends that the relative viscosities of sample and eluent not differ by more than a factor of two, which corresponds to a protein concentration of about 70 mg/ml when the eluent consists of dilute aqueous buffer. The relative viscosity may be easily compared by measuring emptying times from a pipette.

For difficult situations the sample size should be small relative to the volume of the column, typically 1 to 5 percent. Reducing the sample size to less than 1 percent of the total column volume does not improve resolution. In situations where high resolution is obtained and the peaks are well separated, it may be possible to determine a higher maximum load empirically. Because no adsorption takes place between the sample and the matrix, the capacity of a gel filtration column has a quite different meaning from that of an ion exchange or affinity column. The capacity of a gel filtration column is not the amount of product adsorbed; rather it is the maximum amount of sample that can be loaded onto the column without loss of resolution. This volume will depend not just on the pore volume and other characteristics of the matrix but also on the number of contaminants in the sample, as the contaminants (not just the product) also enter the pores. Sample volumes up to 30 percent may typically be used in desalting and buffer–exchange operation. Large sample volumes will minimize dilution during desalting operations.

In practice the sample is concentrated to reduce the volume before it is loaded onto the column. One of the limiting factors for the concentration is the solubility of the protein. As the concentration is increased, the tendency of the protein to precipitate will increase and must be avoided. Also, as the sample becomes more concentrated, interactions among molecules within the sample become more pronounced.

The performance of a given sample in a gel filtration chromatography column also depends on the interactions between the molecules within the sample. The presence of disulfide linkages in proteins may lead to polymerization of protein molecules, in which case reducing agent may be required in the elution buffer to prevent aggregation. Ionic interactions may be broken by the addition of salt (0.05 to 0.2 M), and hydrophobic interactions may be broken by the addition of an organic solvent (2 to 5 percent ethanol).

Some proteins exist as dimers in the concentrated state and as monomers when diluted. When it is desirable to chromatograph these as dimers, the high dilutions that occur in gel filtration are undesirable. The dilution can be avoided through the use of a technique described by Becker (1988), in which the sample is applied in a volume that is large relative to the total column volume. This results in the formation of a plateau region that is essentially constant in concentration.

The flow rate also influences the resolution of gel filtration operations. At high flow rates the smaller molecules will have less time to migrate into the pore of a particle before being swept on to the next particle. At very low flow rates the rate of diffusion of product molecules opposite the direction of flow may become appreciable. Both situations lead to decreased resolution.

Mori (1988), suggests that the decrease in the number of theoretical plates with increasing flow rate is approximately inversely proportional to the square root of the flow rate. Anonymous (1982) suggests that the optimal flow rate for maximum resolution is approximately 2 ml/cm^2 h, although flows up to five times faster may not decrease resolution.

Wall effects play a role in the resolution of narrow–diameter columns, as a result of the faster flow that occurs along the wall because of larger spaces between the column wall and the matrix particles than between the particles themselves. At large diameters the column surface–to–volume ratio becomes small and edge effects insignificant. Manufacturers of some large–scale high–pressure equipment compensate for this with the application of external pressure to a flexible wall, causing the wall to deform around the particle.

14.4 OPTIMIZATION AND SCALE UP

The first step of scale up is to optimize the process at the lab scale. Table 14.2 lists the relevant parameters and some guidelines for their setting. Once these parameters are set, the column is scaled up by increasing only the diameter and sample volume. Holding all other factors constant will assure equivalent resolution at large scale. The column diameter is increased by the square root of the scale–up factor to maintain the same ratio of column volume to the full–scale batch volume. For example, if the lab column were 5 cm in diameter and 100 cm long (volume of 2 l) and the largest sample that gave adequate resolution were 100 ml at a concentration

Table 14.2. Parameters to be set for optimal resolution in gel filtration chromatography.

Parameter	Relevant questions	Rule of thumb
Matrix	Ultimate scale?	Choose incompressible resin for large scale
	Sanitary requirements?	Choose hydroxide resistance for parenterals
Buffer	Ionic interactions?	Add salt
	Hydrophobic interactions?	Add solvent
Flow rate		Start with minimum (5 cm/s) and increase until resolution drops
Sample	Maximum concentration?	High viscosity lowers resolution
	Maximum volume?	Start with 5% of column volume, increase until resolution drops

of 10 mg/ml (batch size of 1 g), a full–scale batch of 100 g (scale–up factor of 100) would have a sample volume of 10 l, the volume of the column would be 200 l, and the dimensions would be 50 cm in diameter and 100 cm long. (With a soft gel a column this large would usually be divided into two or three shorter columns run in series to prevent impeded flow due to compressed packing.)

Scale up in this manner will usually provide resolution equal to the small scale. However, the compressible nature of many gel filtration matrices makes them especially difficult to pack uniformly, and large–column hardware may lead to design limitations that decrease column efficiency. The efficiency of a column is measured by the number of theoretical plates, as described in Chapter 10. The number of theoretical plates is somewhat proportional to the resolution one can expect; typical atmospheric columns may have on the order of a few 1,000 plates, while high–pressure columns have upward of 10,000 plates. Inefficiencies of large columns may require a longer column to give the same number of theoretical plates as the lab–scale column. Alternatively, decreasing the sample size may improve the resolution at full scale.

Because large columns are difficult to pack, care must be taken at this stage. The preferred method is to pack the column under uniform pressure, such as 70 to 280 kPa (10 to 40 psi) for soft gels. Alternatively, Hasko and Bartha (1985) recommend suction pumping, the pump drawing buffer out of the bottom of the column.

REFERENCES

Anonymous, "Gel filtration: theory and practice," Pharmacia Fine Chemicals, Uppsala (1982).

Becker, G. W., "Frontal boundary analysis in size exclusion chromatography of self–associating proteins," in Aqueous Size Exclusion Chromatography, P. L. Dubin (ed.), Elsevier, Amsterdam (1988).

Fischer, L., Gel Filtration Chromatography, 2nd ed., Elsevier, Amsterdam (1980).

Hagel, L. "Pore size distributions," in Aqueous Size–exclusion Chromatography, P. L. Dubin (ed.), Elsevier (1988).

Hasko, F., and K. Bartha, "An easy method to pack large scale stack columns," *Journal of Liquid Chromatography* 8(11):2115–2120 (1985).

Mori, S., "Column efficiency," in Aqueous Size Exclusion Chromatography, P. L. Dubin (ed.), Elsevier, Amsterdam (1988).

FURTHER READING

Basedow, A. M., K. H. Ebert, H. J. Ederer, and E. Fosshag, "Fractionation of polymers by gel permeation chromatography: an experimental and theoretical approach," *Journal of Chromatography* 192:259–274 (1980).

Hjerten, S., "Molecular–sieve chromatography of proteins," in New Techniques in Amino Acid, Peptide, and Protein Analysis, A. Niederwieser and G. Pataki (eds.), Ann Arbor Science Publishers, Ann Arbor, MI (1971).

CHAPTER 15

LARGE–SCALE EXAMPLES

The design of a purification process requires the integration of many alternatives and variables, including product characteristics, unit operations and purification methods, and corporate goals. The complexity of the design process provides an opportunity to gain a competitive advantage to those designers who are familiar with all alternatives and the consequences that follow the selection and use of each. To assist in gaining experience in this area, several examples of large–scale processes are described in this chapter.

As one scans the literature, it becomes readily apparent that the term "large scale" means different things to different people. It is most frequently used to describe the production of quantities that exceed prior levels by tenfold or more. Thus if prior work was limited to microgram production levels, then a few milligrams may be considered by some as large scale. On the other hand, if current practice produces gram quantities, then large scale may refer to kilogram amounts. Because the unit operations involved in kilogram production differ significantly from milligram production, the study of procedures for milligram production may not appreciably increase one's experience for larger, kilogram production. For this reason, the processes delineated here emphasize kilogram and gram scale.

Examples provide helpful vicarious experience for the designer by allowing the evaluation of different steps without the cost of direct testing. Unfortunately, few complete examples have been published. The first four examples presented here have been selected because they contain material balances and have been designed for "large" quantities; they provide information of value to designers even though they may never find commercial application. The first example, a process for the purification of somatotropin, uses a minimum of steps and has good recovery. The second example, a process for the purification of interferon, is more complicated and includes an immunoaffinity step at an early stage. The third example, a process for the purification of dehydrogenase, was designed to be inexpensive as the product is not a pharmaceutical. The fourth example, a process for the purification of tissue plasminogen activator, is a complicated process with poor recovery; it is doubtful whether material made by this process could be licensed for human use. These examples describe process steps, process yields, and an analysis of the design. In addition several other processes are reviewed for the order of the steps but without yield data.

15.1 SOMATOTROPIN (BOVINE GROWTH HORMONE)

Recombinant proteins for use as agricultural products have a market potential that can be measured in tons. Among the candidates for animal use are the growth hormones, which increase meat and milk production. Shahidi (1988) has presented a process for the large–scale production of somatotropin (bovine growth hormone) in a plant with an annual capacity of 10 tons. Though Shahidi specifically dwells on the comparison of batch and continuous operation, his yield and recovery data provide useful instruction.

The process is outlined in Figure 15.1, with yield and purity data summarized in Table 15.1. The protein is overproduced in *E. coli* by fermentation of cells at 30°C with heat induction at 39°C for 4 hours. Fermentation is carried out in two 65,000–l fermenters. The protein forms inclusion bodies that represent 25 percent of total cellular protein, or 2.5 g of somatotropin per liter of fermentation broth. Cells are separated from fermentation broth by centrifugation. Disruption requires two passes through a high–pressure homogenizer at 530 atm. The homogenate is centrifuged and washed with nonionic detergent to remove soluble impurities. Repetition of this step twice gives a sludge containing 20 percent inclusion bodies with an overall yield of 80 percent from the recovery steps.

The next step in the process is separation of the product from other insoluble material, and renaturation. The inclusion–body sludge is treated with a chaotropic agent or detergent (guanidine, urea, SDS) to solubilize the protein. The solution is slowly diluted with buffer and left undisturbed for 36 hours, a time period that is critical for correct formation of disulfide bonds. The total time for the solubilization and renaturation step is 44 hours.

The renatured protein solution is filtered through a 0.2–μm filter to remove particulate material and then concentrated tenfold by ultrafiltration with a 10,000–NMWCO membrane. The concentrate is diafiltered with ten volumes of buffer to remove the chaotrope thoroughly. The combined filtration operations take 18 hours.

The next step is anion exchange chromatography, with the somatotropin adsorbed by the column. The column is washed with equilibration buffer and the protein desorbed by a step gradient.

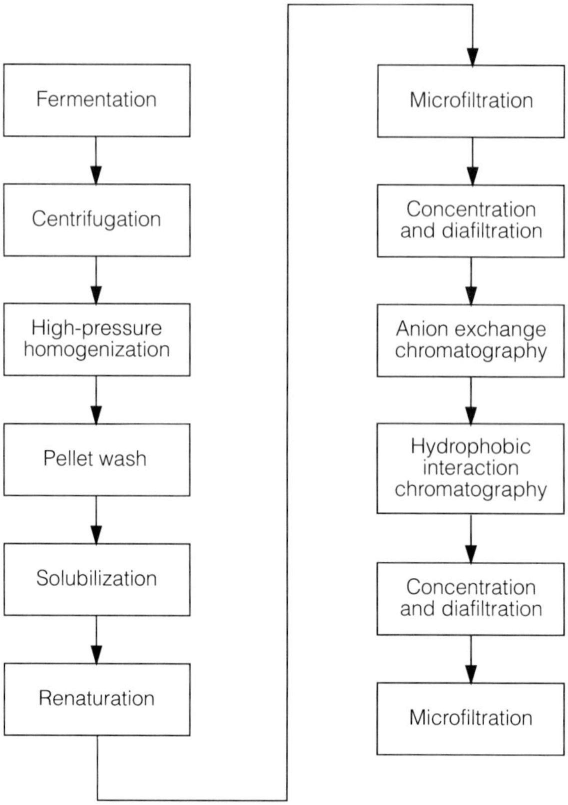

Figure 15.1. Process for purification of somatotropin.

Table 15.1. Yields and purities of steps in purification of somatotropin, starting with 260 kg of inclusion bodies; values are given for the conclusion of each step.*

Step	Volume (liters)	Concentration (mg/ml)	Protein (kg)	Yield (%)	Purity Factor
Renaturation	230,000	0.9	207	80	1.0
Filtration	18,500	9	166	80	1.1
Anion exchange	34,000	3	102	61	3.0
Hydrophobic interaction	15,000	5	75	73	1.1
Filtration	838	80	67	89	1.0
Overall				25	3.6

* Final purity in this case was 98 percent; from Shahidi (1988).

The protein is further purified by hydrophobic interaction chromatography. Following adsorption, the column is washed and the product eluted by a step gradient.

Finally the product is concentrated sixteenfold to a final concentration of 80 mg/ml by ultrafiltration with a 10,000–NMWCO membrane and diafiltered with five volumes of formulation buffer. The solution is filtered through a 0.2–μm microfilter. The overall yield for the process is 25 percent.

Shahidi provides several general observations about the process. Microfiltration was done by cross flow rather than dead–end cartridges because of higher product recoveries from retentate washing. Wide–channel plate–and–frame systems were selected to maximize cross–flow velocity and minimize polarization. For concentration and diafiltration, a spiral–wound ultrafiltration system was chosen because of low hold–up volume, high–pressure tolerance, and low cost.

The performance of the production–scale chromatography steps matched laboratory–scale trials when the bed height was maintained at 25 to 35 cm. The adsorption capacity of the anion exchange resin was 40 mg/ml at a linear velocity of 60 cm/h. The capacity of the hydrophobic resin was 55 mg/ml at a velocity of 90 cm/h. The scaled–up process required 4,200 l of ion exchange resin and 1,900 l of hydrophobic resin. Column volumes were of the order of 400 to 500 l. Shahidi found that the volume of resin could be reduced substantially if the columns were used on a rotating basis, with one being regenerated while another is adsorbing and another eluting, rather than all adsorbing at the same time, eluting at the same time, and regenerating at the same time. Rotating use of columns reduced the volume of resin needed in the ion exchange step to 40 percent of the required volume for strict batch mode and reduced the hydrophobic resin to 50 percent. Semicontinuous operation also reduced the cost of process instrumentation, utilities, tanks, and facilities.

15.2 RECOMBINANT HUMAN LEUKOCYTE INTERFERON (rIFN αA)

Tarnowski et al. (1986) have described the purification of alpha–A interferon from *E. coli*. The process, shown schematically in Figure 15.2, consists of recovery from cells, followed by three column steps: immunoaffinity, cation exchange, and gel filtration. All extraction and column steps were performed at 4°C.

The recovery of the product protein began with the homogenization of frozen cell paste (6 kg) in a meat grinder. The disrupted cells were mixed with extraction buffer (2 M guanidine, 2 percent Triton X–100, volume for volume, 0.1 M Tris, pH 7.5, 24 l) and the pH adjusted to 7.5 with the addition of 5 N NaOH. Following 2 hours stirring, the mixture was diluted with cold distilled water (150 l), which enhanced flocculation of debris and lowered the concentration of guanidine.

Dry filter aid (Hyflow, 40 g/l) was added to the diluted extraction slurry and filtered on a rotary drum vacuum filter that had been precoated with 6 to 10 kg of prepared coating agent (Standard Super Cel). Filtration debris was incinerated. To prevent foaming, 2–octanol was added to the clarified solution, which was further filtered through a 0.3–μm cartridge filter and stored at 4°C. The solution was concentrated threefold by tangential–flow ultrafiltration (10,000 NMWCO) and filtered through a 1.2–μm filter cartridge.

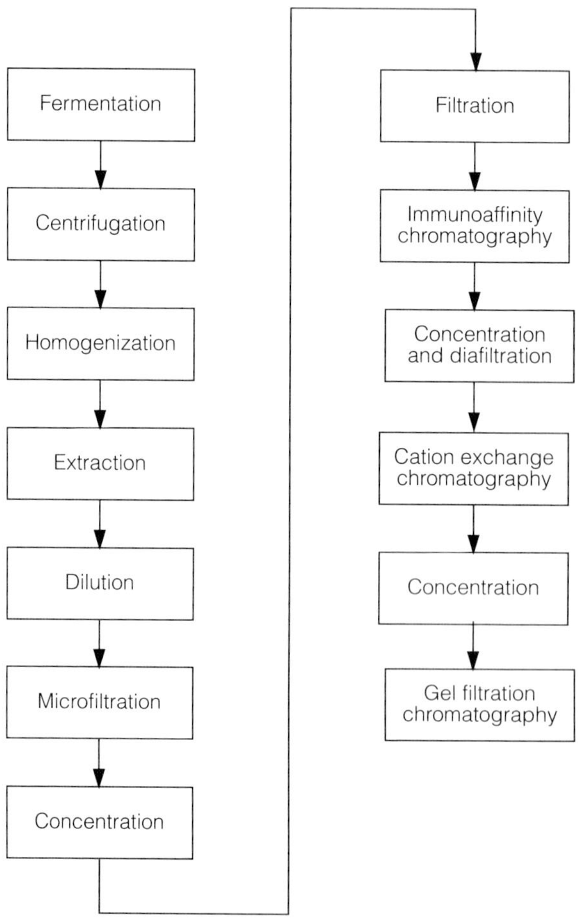

Figure 15.2. Process for purification of alpha-interferon.

The next step was the use of an affinity chromatography column, consisting of monoclonal antibody coupled to N–hydroxysuccinimide–ester–activated Sepharose 4B at a density of 12.7 mg/ml (720 ml total volume). The column was equilibrated with 2.5 column volumes of buffer (0.3 M guanidine, 0.3 percent Triton X–100, 0.1 M Tris, pH 7.5) and the sample loaded at a linear velocity of 38 cm/h. The column was washed with five column volumes of equilibration buffer and each of the following buffers: 0.5 M guanidine, 0.2 percent Triton X–100, 25 mM Tris, pH 7.5; 1 M NaSCN, 0.1 percent Triton X–100, 25 mM Tris, pH 7.5; 0.15 M NaCl, 0.1 percent Triton X–100. The material eluted with these washes was discarded. The product was eluted from the column with buffer (0.2 M acetic acid, 0.1 percent Triton X–100, 0.15 M NaCl) with constant monitoring of pH and ultraviolet absorbance at 280 nm. The peak was collected from the time the effluent decreased below pH 5 until 1.5 column volumes had passed. The column was equilibrated with the starting buffer and another cycle begun. The pool was adjusted to pH 3 to prevent proteolytic degradation during storage.

The accumulated pool was concentrated by ultrafiltration, diafiltered with nine volumes of buffer (25 mM NH_4OAc, pH 4.5, 24 mM NaCl), and further concentrated. The ultrafiltration membranes were washed with the same buffer and the wash added to the concentrated solution. The pool was loaded onto a cation exchange column previously equilibrated with the same buffer used for diafiltration. Under these conditions the product protein bound to the column. The column was washed with equilibration buffer until the absorbance of the effluent measured at 280 nm returned to baseline. As the detergent Triton X–100 absorbs at 280 nm, this indicated that the Triton had been removed from the sample. The adsorbed protein was eluted with buffer (25 mM NH_4OAc, pH 5.0, 0.12 M NaCl) and the peak collected as a single fraction, which was then concentrated by ultrafiltration and filtered through a 0.45–μm filter.

The filtered sample was run on a gel filtration column in the same buffer as was used to elute the cation exchange column. Fractions in the peak representing the monomeric form of the protein were pooled and stored at –20°C. The activity of the monomeric form was approximately three times that of the other peak, which included combined, oligomeric forms of the protein.

The authors made several comments about their process, the yields of which are shown in Table 15.2. They tried centrifugation at 10,000 to 14,000*g* but did not obtain clarity equivalent to the rotary filter. The authors voice concern over the expense of the immunoadsorbent, with regard to the initial material, as well as its life under industrial conditions. The column they employed was subjected to hundreds of cycles; long column life requires that

Table 15.2. Yields and purities of steps in purification of alpha-interferon.*

Step	Volume (liters)	Protein (grams)	Total Activity ($\times 10^{-12}$ units)	Specific Activity ($\times 10^{6}$ units/mg)	Yield (%)	Purity Factor
Crude extract	24	394	2.6	7.0	100	1.0
Drum concentrate	52	192	3.1	16	119	2.3
Affinity eluate	16	8.7	1.9	220	61	13.8
Cation exchange eluate	1	4.9	1.2	250	63	1.1
Gel filtration	0.5	2.4	0.6	250	50	1.0
Overall					23	36

* From Tarnowski et al. (1986).

samples be well clarified before application. The large protein loss on the cation exchange column was due to precipitation of polymeric forms of the product. The cation exchange removed not only the Triton X–100 but was also found to separate the enzymes responsible for proteolytic cleavage. The gel filtration column was effective only at pH 5 and below; at higher pH the protein was polymerized and eluted from the column in the void volume.

This example serves to show how an immunoaffinity step can be used to separate a product protein effectively from most contaminants. The subsequent cation exchange and gel filtration chromatography steps also demonstrate that additives at early steps require subsequent steps for removal and that the hardest separations in biotechnology are between different forms of the same protein.

15.3 L–LEUCINE DEHYDROGENASE

Schutte et al. (1985) have described the purification of L–leucine dehydrogenase, an enzyme that is useful for the synthesis of chiral compounds, such as L–amino acids, by reductive amination of corresponding 2–ketoacids. This is an industrially attractive process for the production of L–leucine and other aliphatic amino acids if the enzyme is available at a sufficiently low cost.

The flow chart for this process is shown in Figure 15.3 and yields and purities are listed in Table 15.3. Homogenization of 30 kg of cells in a bead mill was followed by heating at 63°C for 10 minutes at a pH of 7.5 to 8.5. As an alternative to centrifugation in a tube–type rotor, a two–stage extraction process was developed for recovery of the enzyme. In the first stage, polyethylene glycol 1540 (18 percent, weight for weight) and potassium phosphate salt (7 percent, weight for weight) were added to 70 l of crude extract from the homogenizer, and deionized water was added to a total mass of 140 kg. The polyethylene glycol–rich upper phase, which contained 98 percent of the dehydrogenase, was separated from the solids containing lower phase by a centrifuge with a disk–type rotor. In the

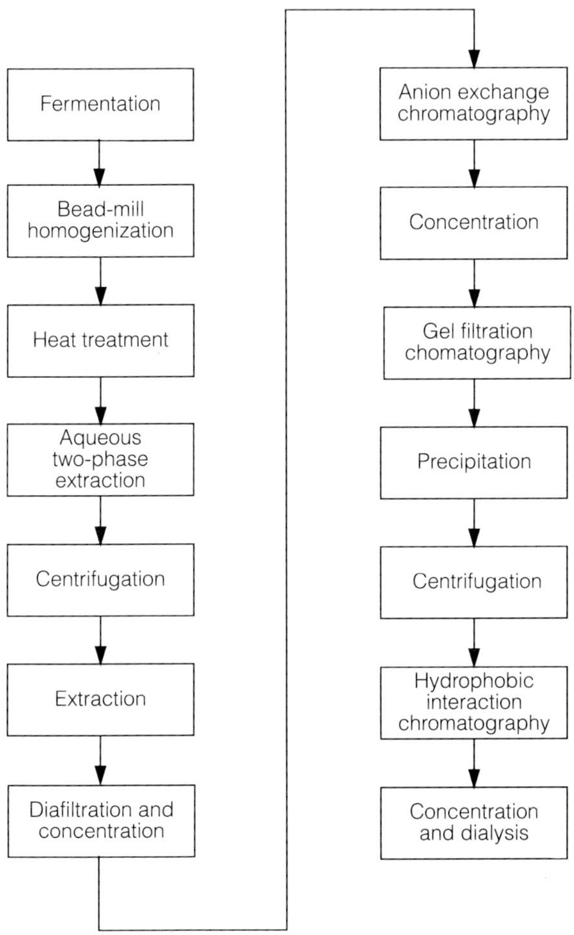

Figure 15.3. Process for purification of L-leucine dehydrogenase.

Table 15.3. Yields and purities of steps in purification of L-leucine dehydrogenase.*

Step	Volume (liters)	Protein (grams)	Total Activity ($\times 10^{-3}$ units)	Specific Activity (units/mg)	Yield (%)	Purity Factor
Crude extract	70	723	434	0.6	100	1
Heat treatment	70	145	421	2.9	97	5
Upper phase I	86	61	412	6.8	98	2.3
Lower phase II	99	40	373	9.3	90	1.4
Diafiltration	2.4	22	360	16	96	1.7
Anion exchange	4.7	13	308	24	85	1.5
Gel filtration	0.48	6.3	260	41	84	1.7
Hydrophobic interaction	1.4	3.6	208	57	80	1.4
Overall					47	95

* From Schutte et al. (1985).

second stage of the extraction, potassium phosphate salt (11 percent, weight for weight) and sodium chloride (to a final concentration of 0.3 M) were added and the mixture stirred for 1 hour. Gravity settling for 30 minutes gave complete phase separation. The lower phase, enriched for dehydrogenase, was diafiltered against buffer (10 mM potassium phosphate, pH 7.5) and concentrated by ultrafiltration.

The mixture was chromatographed on a weakly basic ion exchange resin previously equilibrated with the diafiltration buffer. Following thorough washing with equilibration buffer, the product was eluted with a linear gradient of sodium chloride (0 to 0.25 M) in equilibration buffer. Fractions containing active enzyme were concentrated by ultrafiltration and chromatographed by gel filtration at a linear velocity of 1.1 cm/h.

The final step used a precipitation technique referred to by the authors as "interfacial salting–out chromatography" (but which looks suspiciously like hydrophobic interaction chromatography). The pooled fractions from the gel filtration step were dialyzed against buffer (50 mM potassium phosphate, pH 7.5, containing ammonium sulfate at 35 percent saturation). The precipitate that formed after overnight dialysis was removed by centrifugation and discarded. The supernatant was applied to a column containing unsubstituted agarose gel equilibrated with the dialysis buffer. The dehydrogenase was adsorbed to the column (presumably by hydrophobic interaction), and unbound protein was washed with equilibration buffer. Dehydrogenase was eluted by a decreasing linear gradient of 40 to 20 percent ammonium sulfate in equilibration buffer. Pooled fractions were concentrated by ultrafiltration and dialyzed against buffer (50 mM potassium phosphate, pH 7.5, 0.1 percent 2–mercaptoethanol) and stored at 4°C or with 50 percent glycerol at −20°C.

Though several of the steps used in this process are uncommon, they have been effectively applied. This illustrates how a process designer with a wide range of knowledge concerning alternative unit operations can apply them in ways that provide an advantage. However, before using steps such as interfacial salting–out chromatography, one must to ascertain that it is an appropriate method for the situation, especially with regard to reproducibility and ease of use in the setting for which it is intended.

15.4 TISSUE PLASMINOGEN ACTIVATOR (TPA)

Harakas et al. (1988) have reported the purification of tissue plasminogen activator from cultured human cells. Plasminogen activators, which convert plasminogen to a serine protease known as "plasmin," include streptokinase, urokinase, tissue plasminogen activator, and vascular plasminogen activator. Plasminogen activators dissolve blood clots and are clinically useful in the treatment of heart attack.

In the process described by the authors urokinase and TPA were both secreted by the producing–cell line, and therefore the process needed to separate these two similar molecules. All purification steps were conducted at 4°C, with the exception of the gel filtration step, which was at 22°C. To avoid loss of the hydrophobic TPA molecule onto equipment surfaces, 0.01 percent (volume for volume) Tween 80 was added to the serum–free conditioned media, plastic (hydrophilic) equipment was used wherever possible, and equipment surfaces were prewetted with phosphate buffered saline solution containing 0.01 percent (volume for volume) Tween 80. The flow chart of the process is shown in Figure 15.4; yield data are presented in Table 15.4.

The first step was zinc chelate chromatography using zinc chelate Sepharose 6B charged with Zn^{++} ($ZnCl_2$) and equilibrated with buffer (1 M NaCl, 20 mM Tris, 0.01 percent Tween 80, pH 7.5). The capacity of the resin was 1 l of conditioned media per 35 g of resin. All plasminogen–activator activity was bound to the resin and eluted from the column with buffer

(1 M NaCl, 20 mM Tris, 0.01 percent Tween 80, 50 mM Imidazole, pH 7.5).

The second step was affinity chromatography using Concanavalin A Sepharose 4B (Con A) to resolve urokinase and TPA fractions, presumably on the basis of different degrees of glycosylation of the two proteins. The zinc chelate fraction was applied to a column equilibrated with buffer (1 M NaCl, 10 mM potassium phosphate, 0.01 percent Tween 80, pH 7.5), under which conditions 90 to 95 percent of the urokinase does not bind. The TPA was eluted from the column with a linear gradient from 0 to 100 percent buffer (10 mM potassium phosphate, 0.01 percent Tween 80, 0.4 M α–D–methylmannoside, 2 M KSCN, pH 7.5).

The Con A fractions containing TPA were prepared for gel filtration chromatography by dialysis against ultrapure water containing 0.01 percent Tween 80, lyophilization, dissolution in buffer (20 mM sodium phosphate, pH 6.8, 1.6 M KSCN, 0.01 percent Tween 80) at 4°C, and centrifugation at 6,000*g* for 10 minutes.

The gel filtration chromatography step employed Sprogel TSK 3000 SWG resin in a column of 21.5–mm diameter by 600 mm long at a flow rate of 5 ml/m. The fractionation range of this resin is 15,000 to 150,000 D. The TPA–rich fractions were prepared for the final affinity chromatography step by dialysis against buffer (pure water, 0.01 percent Tween 80), lyophilization, and dissolution in pure water.

The final step consisted of affinity chromatography with a p–aminobenzamidine agarose gel equilibrated with buffer (50 mM Tris, pH 8, 0.5 M NaCl, 0.01 percent Tween 80). Following sample application, the equilibration buffer was used to wash the column. The purified TPA was eluted with elution buffer containing 1 M arginine or 2 M KSCN.

This process is basically the same as the method developed by Rijken and Collen (1981) with the addition of the p–aminobenzamidine column. This example is included because it contains material balances, not because of its ability to produce high–grade product. It is doubtful that this process could produce material of commercial quality in terms of both cost and purity. The Harakas process yielded 20 mg from 1,400 l; the Rijken process yielded 1 mg from 10 l, both with similar activities (110,000 international units of TPA per milligram for Harakas and 90,000 international units of TPA per milligram for Rijken). Builder et al. (1988)—designers of the licensed commercial process—commented that poor yields from natural sources indicate "there is significant doubt that any nonrecombinant cell line would have a high enough level of productivity to be economically viable."

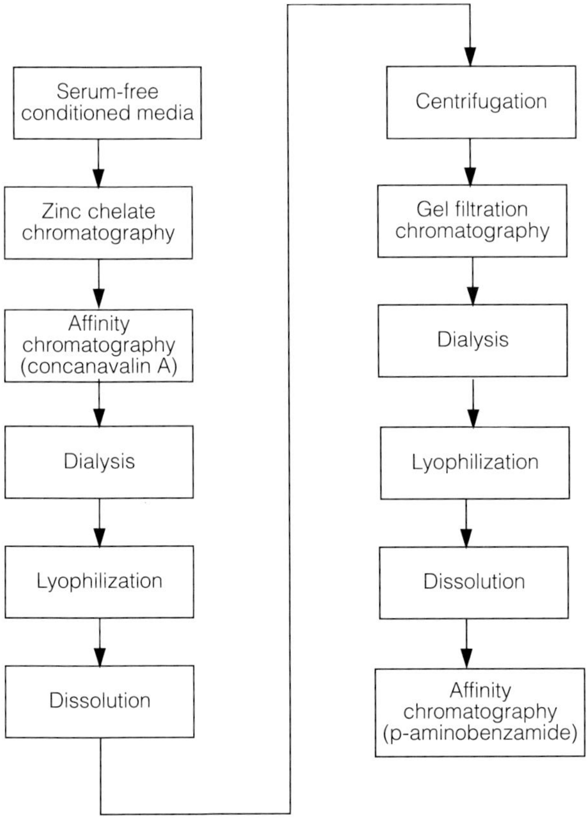

Figure 15.4. Process for purification of tissue plasminogen activator.

Table 15.4. Yields and purities of step in purification of tissue plasminogen activator.*

Step	Volume (liters)	Concentration (mg/ml)	Protein (grams)	Yield (%)	Purity Factor
Media	1400	0.03	46.8	–	1.0
Zinc chelate	41	0.1	4.1	91	12.0
Con A	3.6	0.08	0.3	10	15.0
Gel filtration	0.5	0.16	0.08	84	5.0
p-AB	0.04	0.5	0.02	77	2.0
Overall				6	1800

* From Harakas et al. (1988).

This example emphasizes the point that the most difficult separations are between similar forms of the same protein. Note the considerable treatment required before and after the gel filtration column (dialysis, lyophilization, dissolution). It is not clear whether this is required for the stability of the protein and whether it contributes to the low yield.

15.5 OTHER EXAMPLES

The preceding four examples describe in detail the purification of four different proteins. The processes are marked by distinct preferences of the process designers, and the variations illustrate some of the options available for purification. The yield data indicate the relative difficulty of separating similar species, with the more difficult separations having significantly lower yields. The variation in yields also illustrates the value of high recoveries: A sevenfold difference can have a significant impact on the cost of the final product.

The following examples describe additional processes that have been used for large–scale purification. Unfortunately yield and recovery data are not available, but there are still several lessons that can be learned from them.

15.5.1 Monoclonal Antibodies

Malm (1985) has described a general method for purification of monoclonal antibodies based on ion exchange chromatography. The procedure gives a hundredfold concentration and produces product of 70– to 95–percent purity. The first step is buffer exchange and removal of low–molecular–weight contaminants by desalting. The new buffer is 20 mM sodium citrate, pH 5.3. The second step is cation exchange, with the column equilibrated with the desalting buffer. The antibody is applied to the column, the column is washed with equilibration buffer until the ultraviolet absorbance returns to baseline, and the column is eluted with a step gradient of buffer (20 mM sodium citrate, 0.14 M sodium chloride, pH 5.0). Following elution, the column is cleaned with 0.1 M NaOH. The ion exchange step concentrates the product and provides most of the purification. The third step is a gel filtration column to provide increased purification.

This simple process contains three steps: two based on differences in the size of the product and contaminants and one on differences in charge. No sample treatments are needed between steps as each step leaves the sample in a form ready for application in the next step. This is an important point with regard to yield because each manipulation results in some finite loss of product; so the fewer manipulations, the better. This is the same as playing smart pool: The best shot leaves your cue ball in position for the following shot. The conditions for the ion exchange column described here may not be appropriate for other antibodies within the same subclass or from different subclasses owing to differences in isoelectric point, and further optimization may be needed for different starting material.

15.5.2 Insulin

Prouty (1989) and Kroeff et al. (1989) have described the purification of recombinant insulin as practiced by Eli Lilly and Company. Insulin was the first recombinant protein on the therapeutic market, and its market potential is measured in hundreds of kilograms. As manufactured by Lilly, insulin is produced in *E. coli.* Following fermentation, the cells are chilled and lysed and product–containing inclusion bodies are isolated. The mixture is treated with CNBr to cleave proteins into small fragments, leaving the proinsulin as a large, uncleaved molecule. The S–sulfonate derivative of proinsulin is isolated and chromatographically purified by unspecified steps (probably ion exchange), after which the S–sulfonate is folded to form proinsulin, which is further purified (again by unspecified steps) and enzymatically transformed to biosynthetic human insulin. The crude material is then purified by multiple high–resolution techniques to remove contaminants, which are chemical variants of insulin.

The first of the high–resolution steps is ion exchange chromatography followed by crystallization. The crystals are redissolved for the second step, reversed phase HPLC. The conditions for this step are given by Kroeff et al. (1989). The solid phase was Zorbax Process Grade C_8. The insulin was loaded onto the column in a water–rich mobile phase and was eluted from the column with a linear gradient ranging from 15 to 30 percent acetonitrile at better than 97 percent purity and over 75 percent recovery. A step gradient that gave equivalent purity and recovery was not found, probably because of the nonlinear isotherm exhibited by the product in the overload conditions at which it was run. The column was regenerated with buffer (50 mM ammonium phosphate, pH 7.4, 60 percent acetonitrile). The insulin was recovered from the elution buffer by precipitation with zinc chloride. Further purification was accomplished by gel filtration chromatography. The largest column reported for the reversed phase step measured 50 by 45 cm with an internal volume of 80 l and an insulin load of 1.2 kg.

15.5.3 Protein *A*

Love et al. (1987) have described the production of recombinant protein *A*, which is found in nature as a cell–wall component of *Staphylococcus aureus*, and which is useful in affinity purification of a wide range of antibodies. Their process begins with 20 kg of recombinant *E. coli* from fermentation, which is lysed in a bead mill and clarified by tangential–flow filtration in 0.1–μm hollow fibers. The clarified lysate was heated to 80°C and held there for 30 minutes to precipitate many *E. coli* proteins.

The mixture was again filtered through a cross–flow 0.1–μm filter and diafiltered with a 10,000–NMWCO membrane to reduce the ionic strength and remove low–molecular–weight impurities. The next step was an anion exchange column (45 l in volume, 15–cm bed depth, 160–cm/h linear velocity). The protein *A* was adsorbed to the column equilibrated with buffer (25 mM Tris, pH 8.3) and endotoxin removed by repeated washing with the same buffer containing 1 percent Triton X–100. Detergent was removed from the column by washing with equilibration buffer. The protein *A* was eluted with buffer containing 0.15 M KCl. The last step in purification was a precipitation with 75 percent ethanol at 4°C. The resulting product was 98 percent pure, and endotoxin levels were less than 1 endotoxin unit per milligram. Batches of 250 g were routinely prepared by this protocol.

15.5.4 Enterotoxin *B*

Johansson et al. (1987) have described a process for the purification of enterotoxin *B* from *S. aureus* cell culture supernatant, a product that is of interest in oncology and immunology because it is mitogenic and is a causative agent of food poisoning.

Following fermentation, cells were removed by centrifugation and the supernatant was clarified by filtration. The ionic strength was reduced by dilution with four volumes of water, and glacial acetic acid was added to pH 5.6. A cation exchange column was equilibrated with buffer (50 mM sodium acetate, pH 5.6) and the diluted solution applied at a linear flow rate of 400 cm/h. The column was washed with equilibration buffer and the product eluted with buffer (30 mM phosphate, pH 7.5) at a linear flow rate of 30 cm/h.

The eluted material was adjusted to pH 5.6 and reapplied to the same type of ion exchanger at a linear flow rate of 150 cm/h. Elution was performed with a step increase in ionic strength (0.22 M sodium acetate, pH 5.6) at a linear flow rate of 30 cm/h. The final step was gel filtration chromatography, the sample representing 2 percent of the total column volume and at a linear flow rate of 20 cm/h. The purity of the product was better than 99 percent as determined by Coomassie–stained SDS gels and analytical HPLC with a gel filtration column. The overall recovery was about 45 percent. The use of the first column to concentrate a large volume of culture broth with a fast flow rate illustrates an advantage of ion exchange, namely, the avoidance of extensive sample preparation and rapid adsorption with elution of a concentrated product.

15.6 SUMMARY

These examples serve to illustrate several features of protein purification processes. Perhaps the most apparent is that each of these examples is totally different from the others. We can draw some generalizations from this, namely, that different proteins require different treatments to purify them; the same protein generated by different organisms requires a different purification procedure; and each designer will choose different steps, depending on personal preference. The number of possible alternatives is practically limitless. In this book we have looked at many different bases for separation, including over a dozen affinity methods, multiple ion exchange methods, other chromatographic techniques, and precipitation, extraction, and so forth. Altogether we have described over fifty alternative separation bases; and if we take any three of these for a process, the number of alternative processes we can generate is 50^3, or 125,000. If we take five steps per process, the number jumps to 3×10^8, and if we consider the variations from changing pH, salt concentration, and the other flexible parameters, the number of unique combinations is very large indeed. Within the realm of reality, however, the process designer cannot try all possible options but must be limited to only a few. We have looked at methods of limiting the trial solutions to those most likely to succeed.

We have considered many of the operations available for the separation of proteins. Each of the chromatographic processes was discussed, separately and in general. The widest range of high–resolution steps is found in chromatography, although it can also be used for crude recovery operations. We have considered the unit operations of extraction, precipitation, centrifugation, filtration, and concentration, each of which has a particular advantage in certain situations. The various unit operations exploit many physical and chemical bases for distinguishing among molecules: ionic charge, molecular size, specific binding, hydrophobicity, isoelectric point, density, sedimentation rate, solubility, and so on.

Beyond the actual separation of the product from the impurities lie several considerations that bear on our choice of unit operations, including regulatory concerns, safety, sanitation, equipment limitations, economies of scale, capacity, and corporate philosophy. Limitations are also imposed by the upstream part of the operation: whether the product is secreted and whether it is soluble. We have reviewed how conventional wisdom has been distilled into a handful of rules of thumb, or heuristics, for selecting between alternative steps in the face of the horrendous complexity of all the alternatives taken at once. And beyond this are the basic business considerations of how one allocates scarce resources: The emphasis may be on being first to market, on having the highest quality (or unique advantage) in a niche market, or the lowest cost. And once all this is complete, we meet the test of survival in the marketplace, perhaps the biggest challenge of all.

REFERENCES

Builder, S. E., R. van Reis, N. Paoni, and J. Ogez, "Process development and regulatory approval of tissue–type plasminogen activator," paper presented at 8th International Biotechnology Symposium, Paris (1988).

Harakas, N. K., J. P. Schaumann, D. T. Connolly, A. J. Wittwer, J. V. Olander, and J. Feder, "Large–scale purification of tissue type plasminogen activator from cultured human cells," *Biotechnology Progress* 4(3):149–158 (1988).

Johansson, H. O. J., N. T. Pettersson, and J. H. Berglof, "Development of a process for large–scale purification of enterotoxin B from *S. aureus* cell culture supernatant," paper presented at 3rd Washington Symposium on Preparative Scale Liquid Chromatography, Washington (1987).

Kroeff, E. P., R. A. Owens, E. L. Campbell, R. D. Johnson, and H. I. Marks, "Production scale purification of biosynthetic human insulin by reversed phase high performance liquid chromatography," *Journal of Chromatography* 461:45–61 (1989).

Love, R., A. Profy, B. Kaylos, M. Belew, and W. Herlihy, "Large–scale production of recombinant Protein A and its use in purifying monoclonal antibodies," in Protein Purification: Micro to Macro, R. Burgess (ed.), Alan R. Liss, New York (1987).

Malm, B., "A rapid method for isolation of monoclonal antibodies from large volumes of hybridoma cell culture supernatant," paper presented at Biotech USA, Washington (1985).

Prouty, W. F., "Lessons from production of recombinant insulin," paper presented at Biotech USA, San Francisco (1989).

Rijken, D. C., and D. Collen, "Purification and characterization of the plasminogen activator secreted by human melanoma cells in culture," *Journal of Biological Chemistry* 256:7035–7041 (1981).

Schutte, H., W. Hummel, H. Tsai, and M. R. Kula, "L–leucine dehydrogenase from *Bacillus cereus*: production, large–scale purification and protein characterization," *Applied Microbiology and Biotechnology* 22:306–317 (1985).

Shahidi, A. J., "Design considerations for large–scale purification of proteins," paper presented at Biotech USA, San Francisco (1988).

Tarnowski, S. J., S. K. Roy, R. A. Liptak, D. K. Lee, and R. Y. Ning, "Large–scale purification of recombinant human leukocyte interferons," *Methods in Enzymology* 119:153–165 (1986).

FURTHER READING

Garg, V. K., "Large–scale purification of monoclonal antibodies for therapeutic use," in Targeted Diagnosis and Therapy Series, vol. 3: Targeted Therapeutic Systems, J. D. Rowell et al. (eds.), Marcel Dekker, New York (1989).

McGregor, W. C., "Large–scale isolation and purification of proteins from recombinant *E. coli*," *Annals of the New York Academy of Sciences* 413:231–237 (1983).

Thomson, A. R., "Recent Developments in protein recovery and purification," *Journal of Chemical Technology and Biotechnology* 34B:190–198 (1984).

INDEX